AF293779

Alexander Zech

Nonlinear Traction Control Design for Parallel Hybrid Vehicles

disserta
Verlag

Zech, Alexander: Nonlinear Traction Control Design for Parallel Hybrid Vehicles, Hamburg, disserta Verlag, 2021
Zugl.: Berlin, Technische Universität, Diss., 2021

Buch-ISBN: 978-3-95935-580-3
PDF-eBook-ISBN: 978-3-95935-581-0
Druck/Herstellung: disserta Verlag, Hamburg, 2021
Covermotiv: © pexels.com

Bibliografische Information der Deutschen Nationalbibliothek:
Die Deutsche Nationalbibliothek verzeichnet diese Publikation in der Deutschen Nationalbibliografie; detaillierte bibliografische Daten sind im Internet über http://dnb.d-nb.de abrufbar.

© disserta Verlag, Imprint der Bedey & Thoms Media GmbH
Hermannstal 119k, 22119 Hamburg
http://www.disserta-verlag.de, Hamburg 2021
Printed in Germany

Nonlinear Traction Control Design for Parallel Hybrid Vehicles

vorgelegt von
M.Sc.
Alexander Zech

an der Fakultät V - Verkehrs- und Maschinensysteme
der Technischen Universität Berlin
zur Erlangung des akademischen Grades

Doktor der Ingenieurwissenschaften
- Dr.-Ing. -

genehmigte Dissertation

Promotionsausschuss:

Vorsitzender: Prof. Dr.-Ing. Thomas Richter
Gutachter: Prof. Dr.-Ing. Steffen Müller
Gutachter: Prof. Dr.-Ing. Bernard Bäker

Tag der wissenschaftlichen Aussprache: 30. April 2021

Berlin 2021

To Erika and Wolfgang

Acknowledgements

This thesis was developed and written from 2016 to 2020 during my time as PhD student in the division of functional design in the department of predevelopment powertrain at BMW AG in Munich.

First, I would like to thank my doctoral supervisor Prof. Dr.-Ing. Steffen Müller for important suggestions and advice which contributed to the successful completion of this dissertation. I would like to express my gratefulness to Prof. Dr.-Ing. Bernard Bäker I would like to thank for his interest in my work and for reviewing this doctoral thesis.

Furthermore, I especially would like to thank my supervisor at BMW AG Dr.-Ing. Thomas Eberl for initiating this work and his valuable advice. Due to his great functional insight he was always able to advise me and therefore help lead this research in a promising direction. In many conversations and discussions, he gave me critical feedback and useful advice on a technical and personal level.

Special thanks go to my tandem PhD student Elias Reichensdörfer. His deep theoretical knowledge on control systems greatly contributed to the success of this work. In many meetings we discussed models and traction control designs and their application. I would also like to thank Dr.-Ing. Dirk Odenthal who critically reviewed these designs and made their testing possible.

Additionally, I thank my department, especially Dr.-Ing. Andreas Wilde who supported my ideas and financed the project. Special thanks go to Henning Gehrke and Lorenz Niklas for their support. I would also like to thank the division of prototype development for their assistance during the construction of the test vehicle and also the colleagues of functional design in the department of driving dynamics for useful discussions and assistance during tests.

Special thanks go to Osama Al-Saidi for the constructive collaboration and Matthias Förth for helpful discussions. I would like to thank Guanghui Liu, Wolfgang Degel, Matthias Endl, Patrick Reimund, Carsten Marx, Dominik Berchtold and Linda Hundegger for their contributions. Additionally, I would like to thank Niklas Kattner, Technical University of Munich, for his assistance.

I would also like to thank friends and family who supported me. I especially thank Christin Reinbach for her patience and loving support throughout the creation of this work.

Munich, September 2021   Alexander Zech

Abstract

Speed-based traction control has recently gained attention in academia. To study this control task, first a verified plant model is analyzed showing the significance of the drivetrain stiffness and the resulting torsional oscillations. Based on these results, two nonlinear controllers are developed that can be applied to plants with actuators of different dynamic responses. Both make use of input-output linearization to deal with the pronounced nonlinearity of the tire-road adhesion. Also, both designs address torsional oscillations, either passively by a drivetrain speed observer or actively by damping. Two implementation proposals of the developed controllers to parallel hybrid vehicles are presented. Depending on the resulting actuator response, two control allocation designs are incorporated. Both designs consisting of controller and allocation are implemented to a close-to-production parallel hybrid test vehicle for benchmarking. First, the designs are evaluated on a test bench ensuring a controlled and reproducible environment. Then, both concepts are compared in real world driving conditions on a low friction surface for various maneuvers incorporating the most stressful to the system. This is followed by a discussion of the test results.

Kurzfassung

Drehzahlbasierte Traktionskontrolle stößt vermehrt auf Resonanz im akademischen Umfeld. Zur Untersuchung dieser Regelungsaufgabe wird zunächst ein verifiziertes Streckenmodell analysiert. Dies zeigt den signifikanten Einfluss der Antriebsstrangsteifigkeit und den daraus resultierenden Torsionsschwingungen. Auf Basis dessen werden zwei nichtlineare Regler entwickelt, welche auf Strecken mit unterschiedlich dynamischen Aktuatoren angewandt werden können. Beide nutzen die Eingangs-Ausgangs-Linearisierung, um die ausgeprägte Nichtlinearität im Reifen-Straße-Kontakt zu adressieren. Beide Regler behandeln Torsionsschwingungen entweder passiv durch Beobachtung der Antriebsstrangdrehzahl oder aktiv durch Dämpfung. Zwei Implementierungen der entwickelten Regler in Parallelhybridfahrzeuge werden vorgestellt. Abhängig von der resultierenden Aktuatordynamik werden zwei Algorithmen der Regelungsallokation verwendet. Beide Konzepte bestehend aus Regler und Allokation werden in einem Versuchsträger seriennah integriert und gebenchmarkt. Zunächst werden die Entwürfe auf einem Prüfstand mit kontrollierten und reproduzierbaren Umgebungsbedingungen bewertet. Anschließend werden beide Konzepte im Fahrversuch auf Niedrigreibwert für verschiedene Manöver gegenübergestellt. Abschließend folgt eine Diskussion der Ergebnisse.

Contents

List of Symbols

Greek Letters

Symbol	Unit	Description
α_{pedal}	%	Driving Pedal Position
$\alpha_{pressure}$	°	Angle for Pressure Build Up in Cylinder of ICE
Δ	-	Unit Delta
$\lambda_{1..n}$	-	Eigenvalue
λ_r	-	Reference Wheel Slip
λ_x	-	Longitudinal Wheel Slip
μ	-	Friction Coefficient
μ_{max}	-	Maximum Friction Coefficient
μ_x	-	Longitudinal Friction Coefficient
ρ	$\frac{\text{kg}}{\text{m}^3}$	Density of Air
Ω	-	Set of Control Inputs
ω	$\frac{\text{rad}}{\text{s}}$	Rotational Speed
$\dot{\omega}$	$\frac{\text{rad}}{\text{s}^2}$	Derivative of Rotational Speed
ω_{gc}	$\frac{\text{rad}}{\text{s}}$	Gain Crossover Frequency
ω_r	$\frac{\text{rad}}{\text{s}}$	Reference Rotational Speed of Wheel
ω_{diff}	$\frac{\text{rad}}{\text{s}}$	Reference Rotational Speed of Differential Input
ω_{rim}	$\frac{\text{rad}}{\text{s}}$	Reference Rotational Speed of Rim
ω_{belt}	$\frac{\text{rad}}{\text{s}}$	Reference Rotational Speed of Tire Belt
ω_{ICE}	$\frac{\text{rad}}{\text{s}}$	Rotational Speed of Combustion Engine
$\omega_{ICE,scal}$	$\frac{\text{rad}}{\text{s}}$	Rotational Speed of Combustion Engine scaled to Axle
ω_{EM}	$\frac{\text{rad}}{\text{s}}$	Rotational Speed of Electric Motor
$\omega_{EM,scal}$	$\frac{\text{rad}}{\text{s}}$	Rotational Speed of Electric Motor scaled to Axle
ω_{whl}	$\frac{\text{rad}}{\text{s}}$	Rotational Speed of Driven Wheel
τ_{ign}	s	Time Constant of Ignition Path
τ_{na}	s	Time Constant of Air Path Naturally Aspirated
τ_{turbo}	s	Time Constant of Air Path Turbo
$\tau_{red,slow}$	s	Time Constant Reduced Slow Actuator
$\tau_{red,slow,hyb}$	s	Time Constant Reduced Slow Actuator Hybrid
$\tau_{red,slow,ed}$	s	Time Constant Reduced Slow Actuator Electric
$\tau_{red,fast}$	s	Time Constant Reduced Fast Actuator
$\tau_{red,fast,hyb}$	s	Time Constant Reduced Fast Actuator Hybrid
$\tau_{red,fast,ed}$	s	Time Constant Reduced Fast Actuator Electric
ζ	-	Damping Coefficient of Linear Dynamic System

Latin Letters

Symbol	Unit	Description
$\boldsymbol{A}$	-	System Matrix
A	m^2	Area
a	-	Proportional Gain for IOL
a_x	$\frac{m}{s^2}$	Longitudinal Acceleration
$\boldsymbol{B}$	-	Control Effectiveness Matrix or Input Matrix
B_{pac}	-	Coefficient for Pacejka Model
$\boldsymbol{C}$	-	Output Matrix
C_{pac}	-	Coefficient for Pacejka Model
c_x	-	Drag Coefficient
$c_{tmf,red}$	$\frac{Nm}{rad}$	Reduced Torsional Stiffness Two-Mass Flywheel
$c_{hs,red}$	$\frac{Nm}{rad}$	Reduced Torsional Stiffness Half-Shaft
c_{tmf}	$\frac{Nm}{rad}$	Torsional Stiffness Two-Mass Flywheel
c_{cs}	$\frac{Nm}{rad}$	Torsional Stiffness Cardan Shaft
c_{hs}	$\frac{Nm}{rad}$	Torsional Stiffness Half-Shaft
c_{belt}	$\frac{Nm}{rad}$	Torsional Stiffness Tire Belt
$\boldsymbol{D}$	-	Feed-Through Matrix
d	-	Disturbance
$d_{tmf,red}$	$\frac{Nms}{rad}$	Reduced Damping Coefficient Two-Mass Flywheel
$d_{hs,red}$	$\frac{Nms}{rad}$	Reduced Damping Coefficient Half-Shaft
d_{tmf}	$\frac{Nms}{rad}$	Damping Coefficient Two-Mass Flywheel
d_{cs}	$\frac{Nms}{rad}$	Damping Coefficient Cardan Shaft
d_{hs}	$\frac{Nms}{rad}$	Damping Coefficient Half-Shaft
d_{belt}	$\frac{Nms}{rad}$	Damping Coefficient Tire Belt
E_{pac}	-	Coefficient for Pacejka Model
e	-	Control Error
e_r	$\frac{Nms}{rad}$	Rolling Friction Coefficient
f	Hz	Frequency
F_a	N	Air Resistance
F_e	N	Vertical Force
F_r	N	Rolling Resistance
F_x	N	Longitudinal Force
F_z	N	Total Vertical Force
$F_{z,dynamic}$	N	Dynamic Vertical Force
$F_{z,static}$	N	Static Vertical Force
G	-	Transfer Function
g	$\frac{m}{s^2}$	Gravitational Acceleration
g_m	-	Gain Margin
$\boldsymbol{I}$	-	Eye Matrix
i_{gb}	-	Gear Ratio Gearbox
i_{diff}	-	Gear Ratio Differential
i_{tot}	-	Total Gear Ratio
j	-	Imaginary Unit
J_{ICE}	kgm^2	Rotational Inertia ICE
$J_{ICE,red}$	kgm^2	Rotational Inertia ICE reduced
J_{EM}	kgm^2	Rotational Inertia EM

Symbol	Unit	Description
$J_{EM,red}$	kgm^2	Rotational Inertia EM reduced
J_{diff}	kgm^2	Rotational Inertia Differential
J_{pt}	kgm^2	Powertrain Inertia
J_{eff}	kgm^2	Effective Inertia
J_{veh}	kgm^2	Vehicle Inertia
J_{rim}	kgm^2	Rotational Inertia Rim
$J_{whl,red}$	kgm^2	Rotational Inertia Wheel reduced
J_{belt}	kgm^2	Rotational Inertia Tire Belt
$J_{mot,red}$	kgm^2	Rotational Inertia Motor reduced
$J_{mot,red,hyb}$	kgm^2	Rotational Inertia Motor Hybrid reduced
$J_{mot,red,ed}$	kgm^2	Rotational Inertia Motor Electric reduced
J_{red}	kgm^2	Rotational Inertia Drivetrain reduced
$J_{red,yb}$	kgm^2	Rotational Inertia Drivetrain Hybrid reduced
$J_{red,ed}$	kgm^2	Rotational Inertia Drivetrain Electric reduced
K	-	Proportional Gain
K_{damp}	-	Proportional Gain for Active Damping
L	-	Open Loop Transfer Function
$L_{1..3}$	-	Luenberger Observer Gain
l_f	m	Length Front Axle to COG of Vehicle
l_h	m	Length Road to COG of Vehicle
l_r	m	Length Rear Axle to COG of Vehicle
M_S	-	Maximum Value of Sensitivity Function
m_{veh}	kg	Mass of Vehicle
n	-	Measurement Noise
n_{cyl}	kg	Number of Cylinders
n_{ign}	kg	Number of Ignitions per Cylinder and Rotation
p_m	°	Phase Margin
r	-	Reference Value
r_{whl}	m	Wheel Radius
S	Nm	Sensitivity Function
s	-	Laplace Operator
s_m	-	Stability Margin
T	Nm	Torque
T_{act}	Nm	Actual Torque Combined
T_{driver}	Nm	Torque Request by Driver
T_{ICE}	Nm	Actual Torque Combustion Engine
T_{hs}	Nm	Half-Shaft Torque
T_{dt}	Nm	Drivetrain Torque
T_{EM}	Nm	Actual Torque Electric Motor
T_d	Nm	Desired Torque
$T_{d,ICE}$	Nm	Desired Torque Combustion Engine
$T_{d,ICE,fast}$	Nm	Desired Torque Combustion Engine, Ignition
$T_{d,ICE,slow}$	Nm	Desired Torque Combustion Engine, Air
$T_{d,EM}$	Nm	Desired Torque Electric Motor
T_{max}	Nm	Maximum Torque
T_{lim}	Nm	Torque Limitation
T_{lps}	Nm	Load Point Shifting Torque

Symbol	Unit	Description
ΔT	$\frac{\text{Nm}}{\text{s}}$	Rate Limitation of Torque
t	s	Time
t_{delay}	s	Delay Time
t_{ign}	s	Time Delay of Ignition Path
t_{na}	s	Time Delay of Air Path Naturally Aspirated
t_{settle}	s	Settling TIme
u	Nm	Actuator Input
u_{min}	Nm	Minimum Actuator Input
u_{max}	Nm	Maximum Actuator Input
V	-	Proportional Gain for IOL
v	Nm	Virtual Control Input
v_x	$\frac{\text{m}}{\text{s}}$	Longitudinal Vehicle Velocity
w	-	Control Output
W	-	Weighting Factor
$\boldsymbol{W}$	-	Weighting Matrix
$\boldsymbol{x}$	-	State Vector
x	-	State
y	-	Plant Output

Abbreviations

Abbreviation	Definition
2WD	Two-Wheel Drive
4WD	Four-Wheel Drive
ABS	Anti-Lock Braking System
CAN	Controller Area Network
CA	Control Allocation
COG	Center of Gravity
DCA	Dynamic Control Allocation
DOF	Degree of Freedom
ECU	Electronic Control Unit
CECU	Electronic Control Unit of Combustion Engine
DECU	Electronic Control Unit of Driving Dynamics
EECU	Electronic Control Unit of Electric Motor (Inverter)
ESP	Electronic Stability Program
EM	Electric Motor
GPS	Global Positioning System
HEV	Hybrid Electric Vehicle
HIL	Hardware in the Loop
HV	High Voltage
ICE	Internal Combustion Engine
IOL	Input-Output Linearization
LLC	Low Level Control
MABX	MicroAutoBox
MFC	Model Following Control
MTTE	Maximum Transmissible Torque Estimation
PMSM	Permanent Magnet Synchronous Electric Motor
RMSE	Root Mean Square Error
SMC	Sliding Mode Control
ViL	Vehicle in the Loop

1 Introduction

1.1. Motivation

The history of automobiles reaches back into the late 19th century. Since then many technological advancements have made their way into vehicles. The main objectives of these advancements have been to improve efficiency, comfort, safety, and power. But one circumstance remained unchanged for an extended period: Only the driver had full control over the vehicle behavior. The digital revolution changed matters when mechatronics arrived in the automobile in order to support the driver in situations where safety is at risk. These situations occur mainly due to the nonlinear behavior of the adhesion between the tire and road surface. This results in a counterintuitive vehicle behavior thus reducing its maneuverability.

The first electronic system to arrive in industrial vehicles was the antilock braking system (ABS) which prevents the wheels from blocking due to an excessive braking force. This was soon followed by traction control (TC) which prevents the wheels from spinning freely due to excessive engine torque. Since then industry and academia developed and researched algorithms to achieve higher control performance. As the controlled plant has a highly nonlinear behavior this is not a trivial task. Despite this, the industry mainly implements control structures based on gain scheduled PI controllers. The main benefit of these are their comprehensible characteristics and therefore intuitive tuning [103]. Despite this the tuning process is iterative and therefore tedious due to many characteristic curves and maps involved. Academia on the other hand focusses primarily on advanced algorithms which in general have better performance [23]. On the downside, they require highly skilled engineers for parameterization and in some cases do not meet the requirements set by the industry.

Another trend in the automobile industry is the increasing electrification of vehicles. Starting off as a niche product with small production volumes, now all original equipment manufacturers (OEM)s offer electric vehicles of various technologies for the mass market. This trend was strongly enhanced by legislation in different markets worldwide. The objective is to cut emissions of vehicles that are proven to not only have a high impact on climate change but also, on the physical health of people. To meet the imposed requirements, many manufacturers heavily rely on hybrid electric vehicles (HEV) since they offer characteristics which meet customer needs. This is due to a tradeoff between technologies caused by a redundancy in drive torque generation. With the combustion engine long distances can be travelled without the need for long breaks to charge the battery. The electric motor on the other hand, enables pure electric drive for short distances and can therefore be used for commuting or urban driving. This type of vehicle is thought to be a bridging technology from conventional vehicles to pure electric vehicles.

The combination of both disciplines, traction control designed for HEVs is a new research topic. Only few academic papers have been published on this matter. This calls for deeper and broader research on this subject since it offers high potential regarding control performance. In the following first key terms and definitions will be addressed. From this the need for traction control systems and control allocation (CA) becomes apparent. Then known publications on these topics will be discussed in the State of the Art. The publications will be categorized, and performance results of known works will be stated. Also, concepts to resolve actuator redundancies will be pointed out. Special focus here lies on approaches incorporating actuator dynamics due to the significant difference of the dynamic response of combustion engines and electric motors. With this information the knowledge gap will be identified. Thereafter, research questions will be defined and the objective of this thesis will be outlined. The chapter closes with an outline.

1.1.1. Key Terms

To enable reader comprehension of the basic principles of this work a brief introduction of the two main topics addressed will be presented in the following.

Traction Control

The traction control problem arises due to the nonlinear adhesion properties between tire and road: If the torque exerted upon the wheel exceeds the adhesion limit the tires start to skid. When a vertical load acts upon a pneumatic tire a contact surface is formed. Tangential friction forces are generated across this surface. Two mechanisms define this friction: surface adhesion due to intermolecular bonding between tire and road and hysteresis within the rubber as it slides over the road [43, 98]. Both mechanisms depend on slip which is defined by the ratio of the surface velocities to each other [36]. A general definition is given by Rill [109] valid for acceleration and deceleration:

$$\lambda_x = \frac{\omega_{whl} r_{whl} - v_x}{\max(|\omega_{whl} r_{whl}|, |v_x|)}. \tag{1.1}$$

Equation 1.1 defines the wheel slip for traction. It is dependent on the wheel speed ω_{whl}, the wheel radius r_{whl} and the vehicle longitudinal velocity v_x. The slip values are defined between 0, no torque or forces acting upon the wheel to 1, skidding of the wheel at zero velocity. In other words, as soon as torque or forces act upon the wheel, slip occurs which induces a force accelerating or decelerating the vehicle. Figure 1.1 shows the longitudinal and lateral tractive forces for various road conditions as a function of the longitudinal wheel slip. Longitudinal forces first increase with rising slip values and then reach a distinct maximum before declining again. This shows that excessive wheel slip is undesired. Lateral tire forces strictly decrease for increasing slip values. This results in a loss of maneuverability during cornering. Therefore, slip regulation is required.

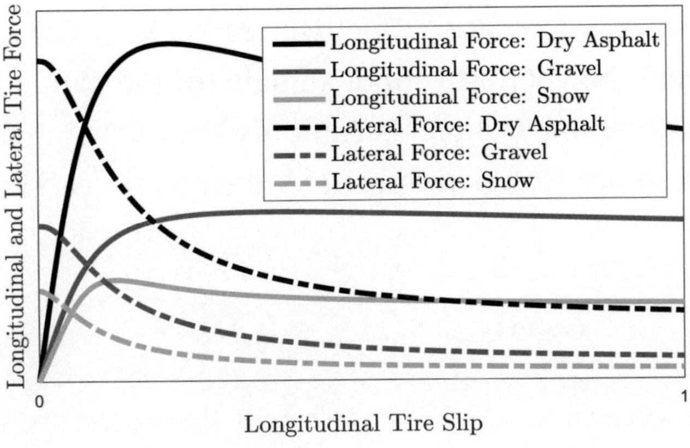

Figure 1.1.: Longitudinal and lateral tire forces as a function of longitudinal wheel slip for various road surfaces - adapted from [98]

Given the information above, the traction control problem can be formulated as defining a feedback system that ensures stable driving conditions during acceleration when too much driving torque is applied by limiting wheel slip.

Control Allocation

Control allocation (CA) finds usage whenever an actuator redundancy is present, see figure 1.2. When using CA the regulation task can be separated from the distribution task [51]. The controller C specifies the total control effort v by evaluating the plant output y and the reference value r. The control allocator CA then determines the distribution u among the available actuators. Both combined are defined as the control system. All actuators then form the desired values u provided by the CA into actual values acting on the plant dynamics P.

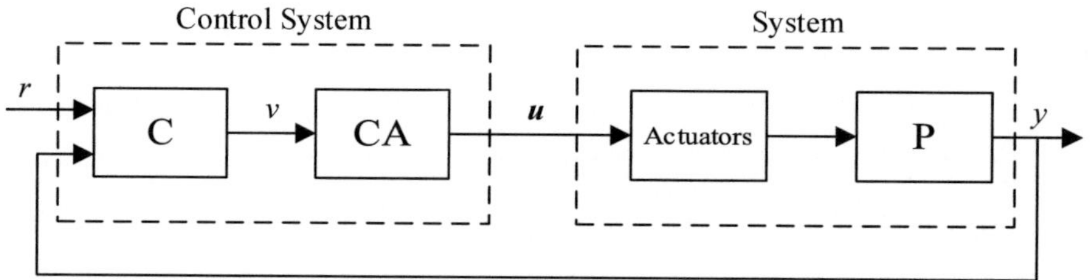

Figure 1.2.: Control Allocation Problem illustrated - adapted from [53]

The CA solves an underdetermined and constrained system of equations. The mathematical description of the linear problem can be formulated as

$$\boldsymbol{B}\boldsymbol{u}(t) = v(t) \tag{1.2}$$

where $\boldsymbol{B}$ is the transposed control effectiveness vector. The control output is described by $v(t)$ and the vector of input signals for the actuators is $\boldsymbol{u}(t)$. This vector underlies further constraints. It can only take values within the boundaries of the actuator. These are determined by the position and rate limits of the actuators

$$\boldsymbol{u}_{min}(t) \leq \boldsymbol{u}(t) \leq \boldsymbol{u}_{max}(t) \tag{1.3}$$

where $\boldsymbol{u}_{min}$ are the lower and $\boldsymbol{u}_{max}$ are the upper limitations of the actuators.

1.2. State of the Art

To derive the objectives of this research, first the state of the art is outlined. Therefore, this section focusses on academic works on traction control and CA methods in the automotive industry.

1.2.1. Traction Control

For a better overview of existing control structures, these will be categorized first. In general controllers can be classified by three main characteristics: the **control input**; the **control method**; and the **control output**. Many works on traction controllers can be classified by these categories. First the latter category will be addressed. Early works [125, 69] deal with throttle control to reduce the torque output of a gasoline engine since the torque was directly controlled through the throttle valve position. This used to be the only relevant actuating variable. Today many additional actuators have been implemented, to enhance power as well as lower fuel consumption. These are for instance turbo charging, fully variable valve, ignition timing etc. [13, 89, 12]. Controlling all at once would make the task unnecessarily complicated. For this reason, the torque-driven-control [42] of automotive motors has been widely implemented. The driver or control systems simply request an output torque making the control task easier. All highly nonlinear actuators are controlled separately to generate the requested torque depending on the operating point as well as satisfying further objectives.

The controls of electric motors in the automotive industry operate similarly enabling the use of scalable functions and controllers across technologies. Again, simply a desired torque is requested on the upper level and underlying controls regulate the current to generate the torque. These controls let the combustion engine as well as the electric motor behave like a linear filter with a time delay, see section 3.3.4. For this reason, only traction control systems that output a torque request are considered for this classification.

Much research on traction control is addressed to electric vehicles since these have been and still are a major trend in the automotive industry. The overall classification presented in the cited research works though may also apply to vehicles driven by combustion engines due to previously described reasons.

Ivanov et al. and Ewin give a clearly presented overview of the state-of-the-art traction controller structures classifying controllers based on the control input on the upper level. The subgroups differentiate between the control methods. Traction controllers are divided into two main groups, **Torque-Based Methods** and **Slip-Based Methods** [36, 65]. The first consists of model following control and maximum transmissible torque estimation. Slip-based methods are divided into two subgroups by Ewin [36], methods without the use of vehicle velocity and direct slip control. However, recently another subgroup of slip-based methods, speed-based slip control, gained attention from the automotive industry [37, 58, 66, 129, 33]. In this thesis slip-based methods will be extended by speed-based slip control methods.

The difference between torque and slip-based methods is that the latter track a desired slip value whereas torque-based methods regulate the slip without a specific setpoint. Therefore a differentiation between the control inputs on the upper level and control method on the lower level has been addressed in this classification. De Pinto et al. sort traction controllers by another characteristic the addressing of elastic driveshafts [103].

1.2.1.1. Torque-Based Traction Control

The first group of traction controllers are torque-based approaches which are further divided into two subgroups. Both do not require an explicit slip reference value. The approaches generate a limiting torque or a torque difference which is subtracted from the requested torque in order to stabilize the vehicle and wheel dynamics. This makes these approaches non reliant on extensive and faulty estimation of the friction curve. On the other hand, the slip cannot be regulated as precisely as with slip-based methods.

Model Following Control

The first approach described in this subgroup is called model following control (MFC). Controllers of this variant of torque-based methods are built on the simplifying assumption that the accelerated inertia drops significantly as soon as the tires slip [5]. The effective inertia J_{eff} can therefore be expressed as [36]:

$$J_{eff} = J_{pt} + m_{veh}r_{whl}^2(1 - \lambda_x) \tag{1.4}$$

For low slip values the inertia consists of the mass of the vehicle m_{veh}, reduced to rotational inertia by the wheel radius r_{whl}, plus the rotational inertia of the drivetrain J_{pt}. As soon as the wheel slip increases, the inertia reduces to the drivetrain inertia only.

The controller is set up as follows. The nominal plant model, which is valid for low slip values, is stimulated with the same input as the plant, see figure 1.3. The difference of their outputs is fed back through a high pass filter and a proportional gain which is then subtracted from the desired torque. This decreases the requested torque as soon as the wheels start to slip.

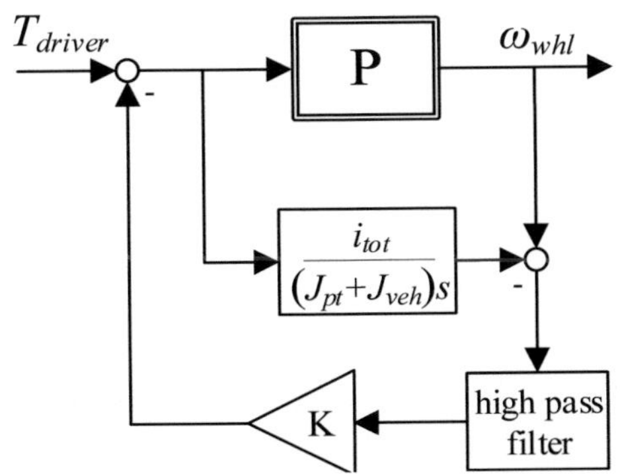

Figure 1.3.: Model Following Control - adapted from [5]

Akiba et al. present experimental results of MFC comparing them to measurements without a controller [5]. The controller prevents the wheels from slipping although high overshoots are observable. More research on MFC was conducted by Hori et al [57] achieving improvements in a test vehicle compared to the non-controlled plant. Further improvements of MFC and model-based traction control have been made by the research group of Hrovat et al. [28] when detailed information about road conditions is limited.

Maximum Transmissible Torque Estimation

Another model-based approach is maximum transmissible torque estimation (MTTE) described in [138]. The basic concept is depicted in figure 1.4 and makes use of a disturbance observer. The plant output is fed back through a low pass and the inverse of the plant model. The plant parameters must be known precisely in order to achieve good control performance. It is then subtracted from the plant input, also filtered by a low pass, and divided by the tire radius. This results in an estimated driving force which is converted to a maximum torque T_{max}. It can be transmitted to accelerate the vehicle. All parameters of this controller are physical values except for one tuning parameter α.

Yin et al. [137] compare the two model-based methods model following control and maximum transmissible torque estimation. This is done through simulations as well as experimental studies. The experiments are conducted on an acrylic sheet lubricated with water to simulate low friction values. The response times and gradients of both controller variants are similar. MTTE outperforms MFC in terms of tracking accuracy. MFC produces a high overshoot, similar to the non-controlled plant whereas MTTE is able to keep the driven wheel speeds significantly closer to the speeds of the non-driven wheel.

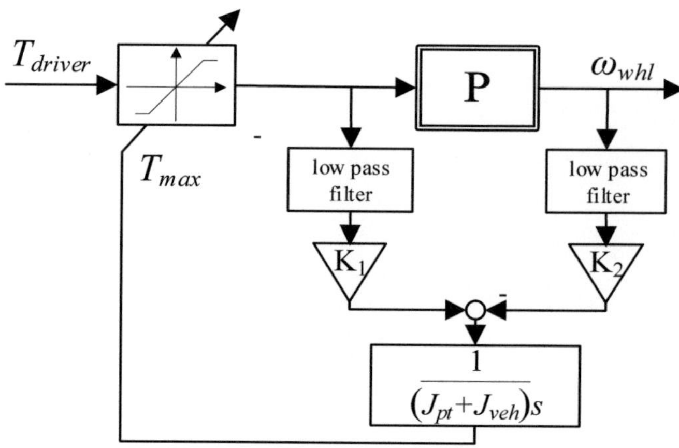

Figure 1.4.: Maximum Transmissible Torque Estimator - adapted from [138]

Hu et al. [59] extend this observer to a closed-loop variant with a feedback loop to compensate for modeling errors. Also, no inversion of the plant model as in classic disturbance observers is needed for the new proposal. Both concepts are benchmarked on a test vehicle which enters a slippery road during high acceleration. Both controllers show similar results although the new approach has a slightly quicker response time and is able to control the driven wheel speeds in a narrower band along the non driven wheel speeds. Also, a study with varying plant parameters for the controller has been conducted showing the robustness of modeling errors.

1.2.1.2. Slip-Based Traction Control

Slip-based approaches further can be divided into three main methods, without the use of vehicle velocity, direct slip control, and speed-based slip control. All methods use a desired slip value, the reference value, which the controller regulates. In slip-based traction control two major difficulties next to the nonlinear control task itself arise: estimation of vehicle velocity and estimation of road conditions to generate the desired slip. The main advantages of these structures on the other hand are, that if these values are known accurately the slip can be regulated very precisely, and traction performance can be increased.

First, the two problems mentioned regarding slip-based control are addressed. Since there are no cost-effective sensors available to measure the velocity under all conditions, various methods are applied. In two-wheel drive (2WD) vehicles the speed sensors of the non-driven wheels yield a good estimate for vehicle velocity but only for longitudinal motion. Sensors that can be used independently from the powertrain configuration are global positioning system (GPS) sensors. Accuracy and availability though limit their usage. Acceleration sensors are another source of a velocity estimate by using its integral. Since it is an open-loop integral, noise and misalignment produce errors so that it can only be used for a limited time, mainly a few seconds. To be applicable to any powertrain configuration, complex sensor fusion approaches are utilized in the automitive industry. Here, uncertain signals from multiple sensors are combined to generate a more accurate estimate of a measured variable. Approaches used for vehicle speed estimation can be found in [46, 7].

Also, the estimation of tire-road adhesion parameters is an important part in slip-based control. TC has a high dependency on the curve-characteristics and therefore, on friction coefficient. A multitude of works deal with its estimation. One example is Stellet et al. [126] where the μ-λ-curve is approximated by a recursive least square parameter estimation of measurement data of μ and λ. Here, the model parameters of the Burkhardt tire model are estimated. Also, Müller presents a method to estimate the road adhesion coefficient using [93]. Based on several input variables regarding weather information a narrow band of possible adhesion coefficients is estimated using a deterministic relationship. Further approaches on the estimation of tire-road friction parameters can be found in [130, 20].

The following TC methods all use an estimated maximum slip value which is the main input to the function. For the first group, this is the only major input variable. Direct and speed-based slip control additionally require information on the vehicle speed.

Slip Control without the Use of Vehicle Velocity

To overcome the reliance on externally provided vehicle velocity information Deur et al. [28] propose a model-based control strategy that estimates the vehicle velocity and the tire forces. A sawtooth signal is added to the force estimation. The control law is a gain scheduled PI-controller. Liang et al. [80] propose a Kalman-like observer which does not use tire-road friction or vehicle parameters but simply relies on acceleration sensors. The integrated error is corrected when the vehicle is coasting. A fuzzy sliding mode controller was chosen.

More advanced studies introduce a slip ratio observer or estimator, see figure 1.5. These were established by Fujii et al. and Fujimoto et al. [39, 40]. The difference in both structures is the use of the longitudinal acceleration in the

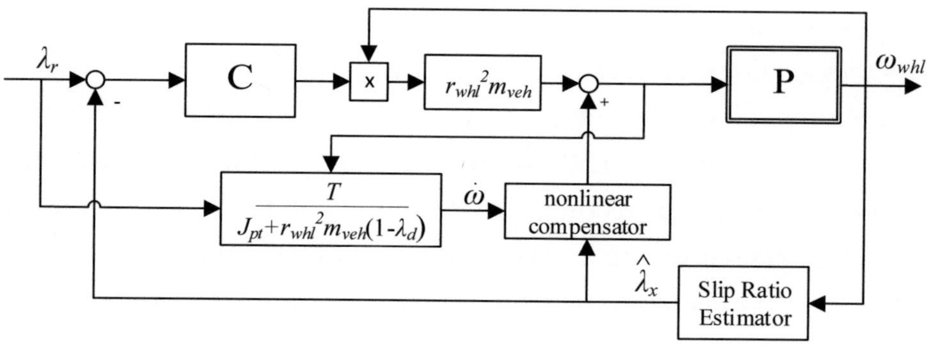

Figure 1.5.: Slip-Based Controller with Estimator - adapted from [39]

observer whereas the estimator does not make use of this state. This is useful to eliminate a steady state error of $\hat{\lambda}_x$. A PI controller as well as one based on feedback linearization is implemented and both are benchmarked. This is performed in simulation as well as in an experimental study with a small test vehicle. Here, the nonlinear controller performance is more accurate due to no steady state error.

Direct Slip Control

This subcategory was treated thoroughly in literature. For this reason, not all controller variations will be discussed but only the most prevalent.

Figure 1.6 shows the general control structure of direct slip control. Depending on the difference of the actual slip and desired slip a limiting torque is generated by the controller C to stabilize the vehicle. There has been much research on how to design this controller. Early works propose a gain scheduled PI-controller [57, 5] where the parameters of the controller are a function of vehicle speed. Due to the highly nonlinear plant dynamics, these controllers lack performance.

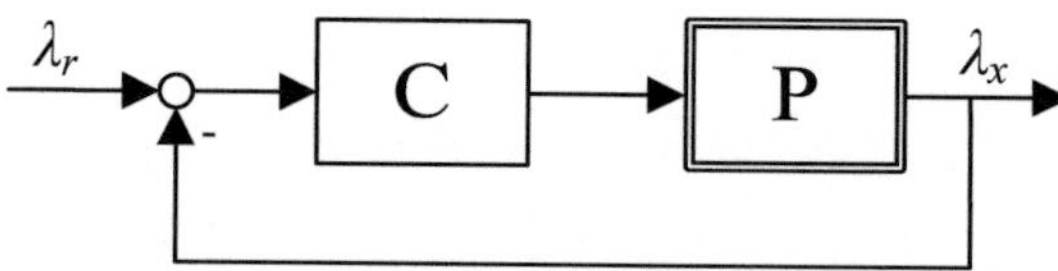

Figure 1.6.: Direct Slip Controller - adapted from [64]

More sophisticated approaches deal with nonlinear control methods such as sliding mode control (SMC) [147, 54]. The objective of such controllers is to reach a sliding surface on which the control error is zero. An advantage of SMC is that it is very robust against uncertain parameters or external disturbances. A major disadvantage is a high frequent chattering of the control output. Two methods that deal with this problem will be described in the following. SMC make use of the sign-function which is the reason for high frequent chattering. Continuous functions, either the tanh-function or a linear curve are a legitimate approximation. They have a smoother progression and no hard cutoffs. Li et al. and de Castro et al. [79, 21] make use of the latter presenting simulation results. Another possibility is to use a higher order of SMC. Hamzah et al. [48] implement a second order SMC presenting a simulation study which highlights the reduction of chattering compared to a first order SMC.

Much research has also been devoted to braking control which has similar dynamics but depends on a different slip definition. Works that also deal with SMC on this topic are [114, 50]. Another approach to the problem is the use of fuzzy controllers [71]. Test bench results are provided showing high frequent chattering of the control action.

Other works on direct slip TC deal with the nonlinear plant dynamics with input-output linearization or feedback linearization. With these approaches the nonlinearity of the system is compensated by interposing its inverse. Therefore, the resulting plant acts linear and a linear controller can be designed. Details on this design method are described in section 2.3.4 and appendix A.1. Fujimoto et al. and Nakakuki et al. [41, 95] propose controllers using feedback linearization. The first research group only presents simulation results but the second extends its studies to experimental results for a sudden change in the road condition from high to low friction coefficient. These show that the feedback linearization controller is able to track the setpoint fast with a low steady state error. Nakakuki emphasizes in the conclusion that the feedback linearization approach offers a simple control structure outperforming a conventional PI controller. Another variant of feedback linearization can be found in [45, 23]. In these works, the P controller is extended by an integral element in order to minimize the steady state error. Guo et al. provide a simulation study arguing the advantages of feedback linearization to be its stability, universality and its good control accuracy under changing road conditions. Chapius et al. compare the feedback linearization controller to a PI and flat controller in an experimental study. Both nonlinear designs are superior to the gain scheduled PI controller which is referred to as the standard in the industry. The flat controller is evaluated to have the best control performance although producing the highest slip overshoot. Regarding complexity and ease of experimental parameter tuning the feedback linearization approach outperforms flat control. Again, many works that deal with feedback linearization in ABS control exist [68, 49, 97] with comparable results to TC.

Speed-Based Slip Control

Speed-based slip control fulfils the same control task as direct slip control but with a different setpoint in the inner cascade, see figure 1.7. In the outer loop, the setpoint is transformed. The new setpoint is driven by the vehicle speed v_x and results in a desired speed ω_r. This makes the behavior of the desired value for the controller change significantly compared to direct slip control. The slip setpoint usually is constant whereas a rotational speed during acceleration gains in value over time which leads to an increasing, slope like character. The inner loop now controls this drivetrain rotational speed which has high dynamics due to the slip characteristics.

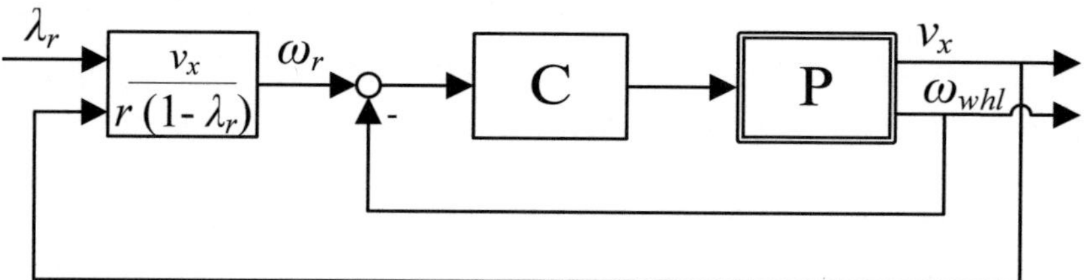

Figure 1.7.: Speed-Based Slip Controller - adapted from [37]

The cascaded structure can be utilized beneficially in industrial applications due to decentralized controls on dedicated electronic control units (ECU) and their delayed communication by bus. Early works on speed-based slip control do not point out this special property or propose a control design [37]. Later, two works made use of speed-based slip control within a different control structure. The research aim of both was set on driving force estimation. For the speed control a PI controller was used [40, 88]. Also, Hrovat et al. propose a PI controller for the speed-based slip control, again without presenting details on the controller design [58].

A different control approach is chosen by Jamie et al. by introducing a PD controller [66]. The control output is added to a direct slip-based controller. Another change was made in the controlled variable. Not the wheel speeds but the rotor speed of the electric motor was chosen as the controlled variable. It was argued that the sensor information is provided faster with this configuration. Syrnik [129] chooses a PID controller with torque limitations to regulate the wheel speed. The described works focus on the benefit of speed-based traction controllers to overall control performance rather than on the controller design itself. An additional functionality next to TC was included in the research project which is rollback prevention. This is discussed for applications in the plane and on an incline. Besides Hrovat et al., all cited publications on speed-based slip control deal with EVs.

1.2.1.3. Oscillation Damping

Further additions to traction control can be achieved by an underlying control algorithm which damps torsional oscillations. This topic was first considered in the pseudolinear region of the μ-λ curve during stable maneuvers [44, 134, 143]. Additions to these dealing with nonlinearities such as backlash were described by Pham et al. [101, 102]. The motivation to damp driveline oscillations arises from the perceived oscillation intensity on humans. For the occurring frequency spectrum of two to six Hz the perceived intensity of humans is high, and reaches its peak, see figure 1.8.

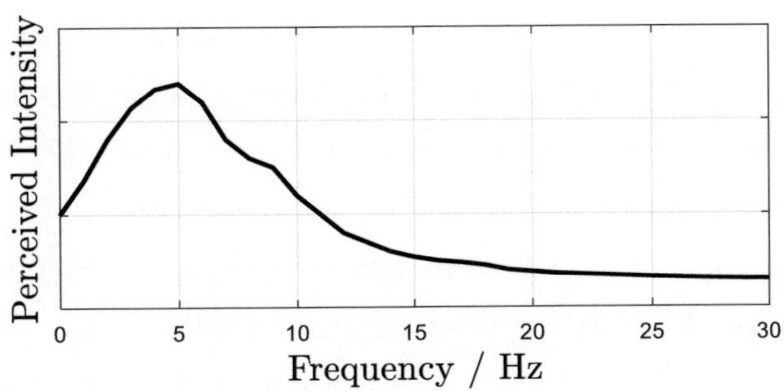

Figure 1.8.: Qualitative Perceived Intensity of Oscillations - adapted from [38]

High torsional oscillations also occur during TC which have a higher frequency spectrum of about 7 to 15 Hz [135]. Dealing with this issue is of high importance to TC as drivetrain torsional oscillations effect not only comfort but also control performance [103]. One approach is to regulate the half-shaft torque with a PID controller [16]. This task though is difficult as the controlled variable is directly dependent on the rotational angles of rotor and wheel. Sensors for these do not exist in industrial vehicles and therefore they must be estimated extensively. For this reason, simple damping controllers have been suggested. Yeap et al. introduces a P controller regulating the difference of wheel and rotor speed [136]. Rosenberger et al. also uses this approach for anti-lock braking control combined with oscillation damping [111]. Another controller has been proposed by Syrnik which shows damping effects for a D controller stimulated by the rotor speed of the EM [129].

A passive approach on damping driveline oscillations during traction control was introduced by Loof et al. [82]. A notch filter set to the resonance frequency was included in the open loop. Quantitative effects of the notch filter approach are not presented.

More additions on traction control consider over-actuated systems such as HEVs or combined brake and engine control. To resolve the actuator redundancy CA is used. Applications of this field in motor vehicles will be described in the next section.

1.2.2. Automotive Applications of Control Allocation

With the emergence of ECUs and integration of new actuators in vehicles CA gained recognition in the automotive industry although having its roots in aviation and marine vessel engineering. An important category of CA is optimization-based CA taking control action dynamics into account. This dynamic control allocation (DCA) can be implemented when actuators have different response dynamics [52]. Various parameters define its behavior. Wang et al. [133] and Härkegard [51] propose DCA for flight control applications. Both authors state that the weighting matrices are of high importance for the control performance but do not give insight in how they were obtained. Zhang et al. propose an addition to DCA incorporating energy consumption [144]. This algorithm is used to address the attitude stabilization problem of a rigid spacecraft. It is compared to a simple control algorithm in a simulation study showing that the DCA approach can reduce settling time and steady state precision of the controlled variables. Information about the parameterization of the weighting matrices is not provided.

CA in the automotive industry finds usage in the two major fields. First there is vehicle stability control due to the abundance of actuators which can apply torque to wheels. The second is HEVs. In the following the state of the art of automotive applications will be described.

1.2.2.1. Vehicle Stability Control

Rajamani classifies three types of vehicle stability control [104]:

1. Differential Braking

2. Steer-by-Wire

3. Active Torque Distribution

Systems of the first category use hydraulic brakes to apply different torques between left and right wheels and therefore control the yaw moment. Known publications on this matter investigate this subject with combined traction control [41, 84, 72]. All concepts make use of two separate controllers regulating traction and yaw moment independently. Steer-by-Wire systems override the steering angle request of the driver to stabilize the vehicle motion. This category only makes use of one single actuator and therefore will not be discussed further. Active torque distribution systems apply torque individually to actuators present in the vehicle to fulfil the control task. These are individual wheel brakes, steering angles, engine torque and active suspension depending on the setup of the vehicle.

State-of-the-art vehicles are equipped with an electronic stability program (ESP) [35]. This enables braking of individual wheels and correcting of the steering angle in order to stabilize the vehicle in critical situations such as oversteering and understeering during cornering. Much research has been conducted on vehicles with four built in electric motors propelling each wheel independently. With this variant also positive torque can be distributed to each wheel compared to systems that only address friction brakes. Park et al. [100] propose a fuzzy control method for torque distribution for a four-wheel drive (4WD) individually driven electric vehicle. The CA is set to maximize energy efficiency. A requested torque is allocated to the four in-wheel motors with regards to keeping losses as low as possible running the motors at the point of highest efficiency. A simulation study is provided confirming higher energy efficiency compared to a fixed distribution. Leng et al. compares static CA approaches of different complexity for vehicle stability control [78]. A simulation study as well as experimental tests are provided. The results show that the most complex, a quadratic problem formulation, has the best precision despite a high control effort. Subroto proposes DCA to distribute torque among all four EMs and control the front and rear steering angle [128]. A simulation study is presented for two maneuvers showing good tracking performance of the proposed concept. The parameterization is not addressed in the contribution.

A general approach for vehicle dynamics control is introduced by Knobel et al. [73, 74]. A nonlinear optimization of the driving forces acting upon the vehicle are allocated using a DCA. The approach can actuate individual brake torques, the steering angle and engine torque. Also, the availability of actuators can be changed online reacting to failures. A simulation study is provided for a medium friction value and a lane change while decelerating. The approach can be used in the early development process to evaluate the potential of different configurations.

1.2.2.2. Torque Distribution in Hybrid Vehicles

Various structures and types of HEVs exist. Torque control of parallel HEVs which this work focusses on can be classified into two main categories, operation strategy, and torque distribution, see figure 1.9.

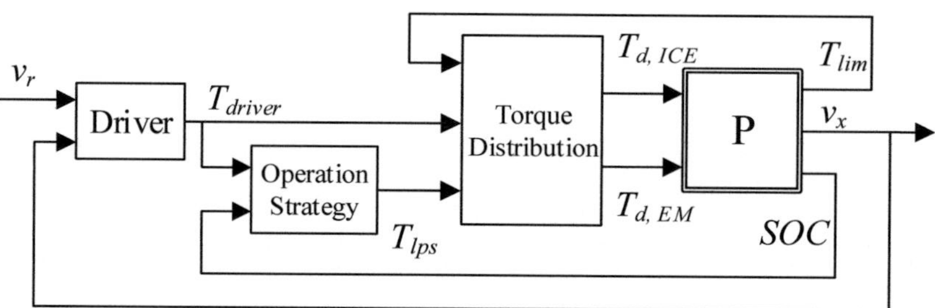

Figure 1.9.: Torque Control of HEVs - adapted from [25]

The operation strategy represents the energy management of the high voltage (HV) system. For this a setpoint for the EM or load point shifting torque T_{lps} is calculated which can take on positive (discharging) and negative values (charging). It is based on the drivers torque request T_{driver} and the current state of charge SOC. In order to still provide the drivers torque request the load point of the ICE is then raised by the EM setpoint. A multitude of different approaches defining the operating point of the EM are discussed in the literature ranging from heuristic strategies to optimization-based strategies [107].

The torque distribution defines the desired torques for ICE $T_{d,ICE}$ and EM $T_{d,EM}$ based on the EM setpoint, and drivers request taking the dynamics of the actuators into account T_{lim}. Only few works discuss the torque distribution in HEVs. One concept is centered on the daisy chain method [132, 56] which is a static allocation method, see chapter 2.4.1. A dynamic allocation method was proposed by Zaccarian et al. [139, 26]. It is a model-based approach using linear models of the torque generation of both actuators. The allocator acts as an underlying controller regulating the difference between the desired torque of the actuators and their individual steady state requests. This ensures that the steady state request of each individual actuator will be met in the long term. Also, the dynamically requested driver torque is satisfied. The overall behavior of the allocator is that of a frequency-separating filter. High dynamic parts are shifted to the fast actuator whereas the low dynamic parts are shifted to the slower actuator.

Further research on HEVs using allocation algorithms address active damping. Looman develops several control approaches to reduce torsional oscillations of the drivetrain since these influence comfort [83]. The work focusses on different states and state changes of the hybrid powertrain rather than on driving dynamics. Only maneuvers in the pseudolinear region of the μ-λ curve are investigated. The controllers each address several actuators in the drivetrain which lets the author make use of DCA among other control methods. A simulation study of the control performance of all approaches is provided. The DCA shows satisfactory performance while being easy to implement and to parameterize. Again, the author does not specify the exact parameters of the used allocation algorithm nor is the method of calibration described.

1.2.3. Slip Control of Over-Actuated Systems

After looking at the two topics, TC and CA, separately the focus now will be set on works that combine both. First, hybrid ABS control concepts will be addressed as this subject is closely related to TC. Then, decentralized solutions that make use of separate controllers each controlling one individual actuator will be presented. Thereafter, centralized solutions which make use of one central controller and an allocation algorithm dividing the control effort among the actuators will be presented.

1.2.3.1. ABS Control of Electric Vehicles

Hybrid concepts for braking control are a particular focus of the academic community as well as the industry. These concepts use an electric motor and the brake system to perform the control task. This new research topic arose with the emergence of EMs in vehicles since ICEs are only able to produce little negative torque whereas an EM can do this by acting as a generator. This is another way to reduce torque upon the wheels to decelerate the rotational motion. Existing slip control methods can still be applied. To address both actuators and to resolve the redundancy a control allocator is introduced. The control allocator assigns the total torque demand upon all actuators by fulfilling underlying objectives.

A method for ABS control is described by Anwar using the daisy chain method to allocate the torque request [8]. All torque is distributed to the EM to slow down the vehicle. If the demand is higher than the EM can supply excess torque is shifted to the hydraulic system. An experimental study is provided by the author. Brake tests on snow were conducted showing good tracking performance of the overall system. Also, full steerability was given during ABS braking for a lane change maneuver.

Rosenberger et al. conducted broader research on this topic. He uses two different approaches to distribute the torque between brakes and EM. The first is a state machine to allocate the torque request [112]. Dynamic torque is allocated to one single actuator while the other is kept at a constant torque. This is done until one actuator reaches its limits in which case the other actuator is excited with the excess torque. This method combines the daisy chain approach with a superordinate controller. The other approach makes use of the DCA [110]. Torque is distributed to both actuators not only based on the current state of information but also on the change from the last control action. With this approach actuator dynamics can be considered explicitly. This method was also used by Tjonnas to regulate wheel slip by actuating brakes and an EM. Additionally Tjonnas adds theoretical remarks of closed-loop control systems using DCA and proves Lyapunov Stability for these. A simulation-based case study is provided. De Castro et al. [22] also uses the DCA to distribute the torque demand of the wheel slip controller to the EM and hydraulic brakes. All three authors neither mention the parameterization method for the DCA nor the chosen parameterization.

Satzger et al. extend the DCA by a model predictive method [113]. Simulation results are presented comparing the DCA approach to its model predictive version. The study shows that control performance of both methods remains the same. The desired slip ratio can be controlled precisely and with similar dynamics upon steps in the reference value. The model-based approach on the other hand is able to meet secondary objectives to a higher degree. The one chosen in the study was the maximization of recuperation. Again, a description of the allocation algorithm parameterization is left out by the authors.

1.2.3.2. Decentralized Traction Control of Over-Actuated Systems

Existing concepts combine the usage of the ICE and the hydraulic brake system in conventional vehicles to regulate the wheel slip during acceleration. Also, HEVs are considered combining the control of ICE and EM. The decentralized approach makes use of independent controllers dealing with the actuator redundancy. This though is not subject to the field of CA. It will be listed here with regards to content.

Reif benchmarks different control outputs for traction control of a conventional vehicle [107]. Three concepts are compared. The first two are engine only control variants where the control output is either the air path or the ignition path of an ICE. The third concept is a hybrid approach, combining ignition path and hydraulic brake control. A comparison of all three concepts is performed qualitatively. The hybrid approach achieves the best result regarding response and settling time. Information on the control algorithm of ICE and brakes is not presented.

A more detailed approach is postulated by Abdelhameed et al. using two separate controllers, a PID controller to regulate the engine torque and a fuzzy controller that actuates the brakes [2]. Both controllers react to the difference in slip ratio. A simulation study is provided for a step in friction coefficient from high to low. Both actuators are stimulated with similar dynamics. Neither the motivation for this nor an exact controller design is mentioned.

Shoubo et al. propose two TC designs for parallel[1] HEVs [123, 24]. Both make use of two independent controllers for EM and ICE. The first design uses a fuzzy PID and the second a linear PID controller to regulate the torque for the EM. The ICE is controlled by a feed forward controller for both designs. Its torque is determined based on the estimated friction coefficient. Experimental results are presented comparing both designs. The tracking performance for the fuzzy PID controller is more accurate than that of the simple PID controller. In general, the ICE shows a low dynamic stimulation whereas the EM is controlled with higher dynamics. The tuning of the PID controllers in this study is

---

[1]See section 2.1 for further information on HEV types

performed by a subjective method as proper gain values are determined by simulation. No objective criteria are mentioned. Additionally, the proposed system causes low frequency torque modulations which are not further investigated.

1.2.3.3. Centralized Traction Control of Parallel Hybrid Vehicles

The centralized approach makes use of a single wheel slip controller combined with a control allocator to distribute torque among ICE and EM. Zhao et al. propose a traction control design also dealing with μ-split road conditions. For this, four actuators are used, ICE, EM and hydraulic brakes individually on the left and right side of the vehicle. First, a multiple input multiple output SMC regulates desired torques on the left $T_{d,whl,l}$ and right side $T_{d,whl,r}$ of the vehicle, see figure 1.10. Then, a coordination centered on an optimization with rule-based constraints defines the desired torque for the hydraulic brakes on the left and right $T_{d,brake\,l/r}$ and the powertrain T_d. The desired powertrain torque is then filtered by a low pass resulting in the desired torque for the ICE $T_{d,ICE}$. The desired torque for the EM $T_{d,EM}$ is determined by an underlying feedback control loop. It regulates the torque difference of T_d and the actual torque of the system.

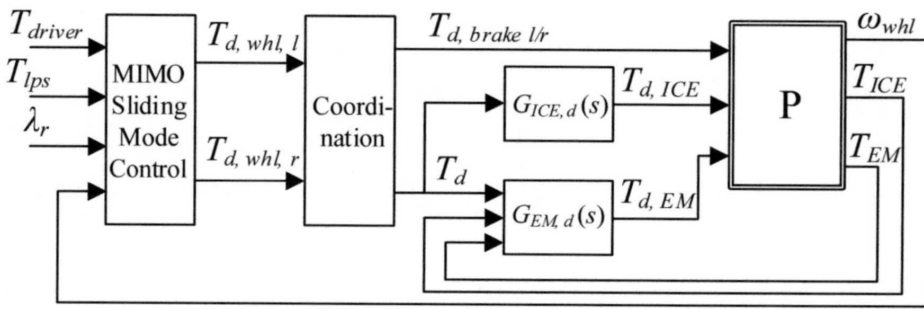

Figure 1.10.: Torque Distribution in HEVs - adapted from [146]

The authors present results of a simulation using a 15-DOF vehicle model as well as measurements of a hardware in the loop (HIL) test bench. The simulation shows good steady state tracking performance during μ-split acceleration. Low damping of the system can be observed due to oscillations during control begin. The measurements of the HIL for the same maneuver show a high initial slip overshoot, compare figure 1.11. Additionally low frequency oscillations of two *Hz* next to higher frequency oscillations occur. Further discussion of the results will follow in the next section.

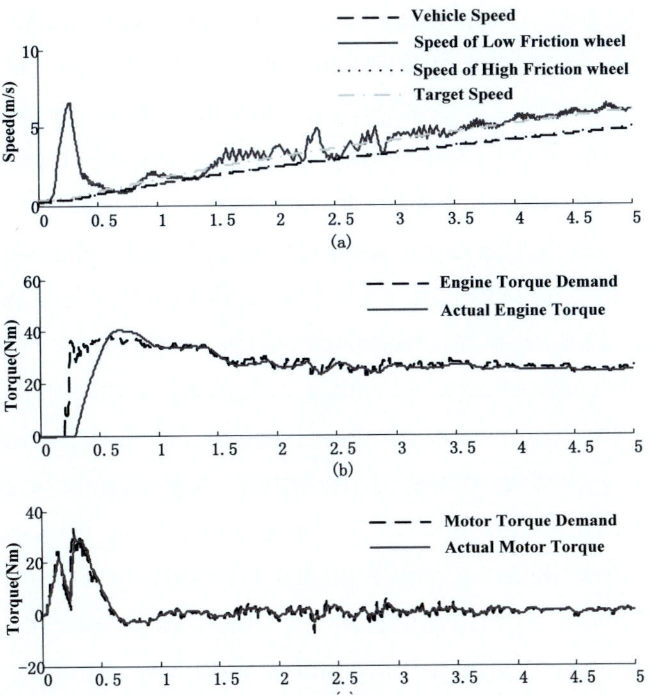

Figure 1.11.: Measurement Results of HIL Test Bench for μ-split Maneuver [146]

Initially there is a high slip overshoot. As the wheel speed overshoots the EM reduces torque at $0.2s$. The ICE can only produce the requested torque with a time delay and low dynamics. The EM compensates this by raising its torque again at $0.25s$. This leads to a high wheel speed overshoot. r performance of the proposed concept compared to ICE and brake torque regulation only.

1.2.4. Summary and Evaluation

The importance of control performance in TC has been pointed out due to its relevance to vehicle stability and therefore, safety. Simple concepts have been described that make use of a reference model to limit the torque of the motor. These torque-based methods only yield a rough control performance. Slip-based methods track a desired reference value and therefore, control wheel slip more accurately. Various nonlinear designs have been proposed which in general show better control performance than linear PID controllers.

Recently, a new control structure for TC gained attention, speed-based slip control. It makes use of a cascade separating low and high dynamic tasks of the control problem. This is a beneficial characteristic for the use in automotive applications due to the communication delays between ECUs. With the cascaded structure low dynamic signals can be transferred by bus generating low phase delays compared to direct slip control where high dynamic signals must be transferred creating high phase delays. The cascaded structure therefore offers better control performance in terms of accuracy and response time. The crucial part for control is the inner loop of the cascade. Known authors though either do not specify this controller [37] or make use of linear PID controllers [40, 88, 66, 129]. The high nonlinearity of the controlled plant limits control performance due to conservative parameterization. Nonlinear designs have not been proposed for the inner loop of speed-based TC. Given the current state of the art, high tracking performance of the inner loop cannot be accomplished.

Torsional oscillations were shown to have significant impact on control performance and comfort. Known publications concerning TC only consider EVs. In contrast ICEs have a significantly lower transient response affecting control performance. No known concept for systems with low actuator dynamics which deals with drivetrain oscillations during TC exists. In general, TC concepts do not incorporate actuator dynamics [103]. Given the current state of the art, TC of systems with low actuator dynamics cannot be carried out with high performance.

Only few works deal with TC of HEVs. Most authors propose two separate controllers dealing with the torque redundancy [123, 24, 2]. This solution though does not meet the requirements set by the automotive industry as it lacks scalability across technologies, see section 4.1. Only Zhao et al. makes use of a central controller [146]. A unique CA method was introduced that takes actual values of the actuators into account. This is a design flaw which can be observed in the provided measurement, see figure 1.11 as it causes high slip overshoots.

The main benefit of HEVs regarding TC in contrast to conventional vehicles is the use of an EM due to its dynamic response. This though can only be harnessed if the integration of the controller within the distributed ECU architecture is chosen beneficial considering communication delays between and operation frequencies of ECUs. None of the known works on TC address this topic. Given this state of the art, TC of HEVs cannot be carried out with good tracking performance.

CA is widely used in the automotive industry with most concepts being static distributors of control effort. This might be reasonable if the control action dynamics are much slower than the actuator dynamics. Only few concepts are known that divide the control effort into distinct parts of the frequency spectrum. A general approach is DCA. This concept has been applied successfully to ABS control regulating hydraulic brakes and an EM [110, 113]. Further applications to non-automotive engineering disciplines have been mentioned. Many authors using the DCA do not specify its parameterization [51, 133]. Most works using DCA describe it as a useful tool to allocate control effort incorporating actuator dynamics although stating that parameterization is of high importance. Here the authors refer to expert knowledge. Given this current state of the art a generic and reproducible parameterization of DCA cannot be carried out. For this reason, further research into the topic of parameterization is of interest. Additionally, only dynamic properties of the CA module is discussed in known publications. The input-output behavior including the actuator dynamics is not investigated. For control design though, this combined system of CA and actuators is of interest.

The cited studies mostly tested the proposed controller concepts in simulation or in a test bench environment offering only limited insights to the performance in real driving conditions. Experimental studies in industrial vehicles though is essential to prove its functionality in applications as simulation models and HiLs are a simplification.

1.3. Objectives

The state of the art of TC and CA in HEVs as well as gaps in the current research have been pointed out. Additionally, objectives that cannot be achieved with the current state of the art were described. This information is now used to formulate the central research question of this work.

Research Question: *To which extent does the utilization of a speed-based controller designed for fast actuators increase performance measures of traction control of HEVs?*

To answer the research question, the following objectives are defined. Actuator dynamics in TC have not been considered in literature so this topic will be addressed in this work. Therefore, two controllers for the speed-based traction controller will be designed. The first design finds application in actuators with low dynamic responses such as ICEs. The second will be applied to actuators with high dynamic responses such as EMs. Furthermore, both designs must deal with torsional dynamics of the drivetrain as this has been pointed out to be essential for the performance of traction controllers. In order to be applied to industrial vehicles requirements of the industry must be met.

Objective 1: *The main aim is the design of two nonlinear drivetrain speed controllers for TC that meet the requirements of the automotive industry, taking different actuator dynamics into account, and dealing with torsional oscillations.*

To reach the first objective CA needs to be implemented to solve the actuator redundancy. DCA is a promising allocation method for systems with different actuator dynamics. As there are no publications known to the author that deal with the parameterization of DCA a method is to be designed in the present work. This ensures objectivity and reproducibility for further uses of this allocation approach. Another objective is to ensure a CA system, composed of the DCA and the actuators, that has a non-oscillating, low pass character. This is of high importance for control performance as the controller expects an actuator with this dynamic response.

Objective 2: *Design of an objective parameterization method for control allocation applications and ensuring a low pass behavior of the CA system.*

Testing a proposed concept in a simulation study does not provide a high level of validity. There are several influences even sophisticated models do not depict. Also, diagnostic functions can limit the functionality of newly developed controllers. For this reason, the development of a new function needs to be performed in vehicle tests. Only this ensures proper behavior in a overly complex and cross-linked system such as the software in an industrial automobile.

Objective 3: *Integration of the designed traction controllers and allocation methods to a parallel HEV prototype and benchmarking of the approaches through an extensive test procedure.*

1.4. Outline

The outline of the work is depicted in figure 1.12. To begin with a, brief introduction to the topic has been depicted followed by the motivation of the work. Then, the general problem of TC and CA has been identified. Given this understanding, the current state of the art has been discussed. The focus here has been on TC concepts. Various approaches have been categorized and described, while mentioning their advantages and disadvantages. Another topic was CA in automotive applications with special focus on the parameterization of the DCA algorithm. With this information the existing knowledge gap and objectives that cannot be achieved have been pointed out. The first chapter closed with the formulation of a central research question and three objectives this work addresses.

The next chapter deals with fundamentals required by this work. First, HEVs are categorized and described. An introduction to linear control theory follows. This is extended by the nonlinear design method input-output linearization. Additionally, bus systems and torque control of industrial vehicles are outlined. The chapter ends with an introduction to the theory of control allocation.

In chapter 3 a simulation model of the vehicle will be derived. Special focus will be set on drivetrain and actuator dynamics. Unknown parameters are identified. A verification of the entire vehicle model will be presented. An open-loop verification of the model is depicted by a comparison of the model and measurements reacting to the same inputs.

Chapter 4 deals with the design of two speed-based traction controllers. The first can be applied to systems with a low dynamic actuator response. A second approach will be discussed with additional features that can be applied to highly dynamic actuators. Both designs deal with the presence of torsional drivetrain oscillations uniquely. The chapter closes with a comparison of the designs in a simulation.

First, two concepts to implement the designed controllers in HEVs are described in chapter 5. Then, two allocation concepts are proposed for both concepts. Additionally, an approach to design the parameters of the DCA will be presented and applied for the torque demand allocation in HEVs during TC. Furthermore, the dynamic behavior of the CA system including actuators is analyzed. A simulation study is provided comparing both proposed concepts.

Chapter 6 deals with the application of the proposed concepts to a test vehicle. First, maneuvers and performance measures will be defined before describing the measurement setup. Then, several longitudinal maneuvers with a test vehicle will be presented. Both concepts are benchmarked for hybrid drive.

This work closes with chapter 7 presenting a summary. Finally, a brief outlook on future research regarding the discussed topic is given.

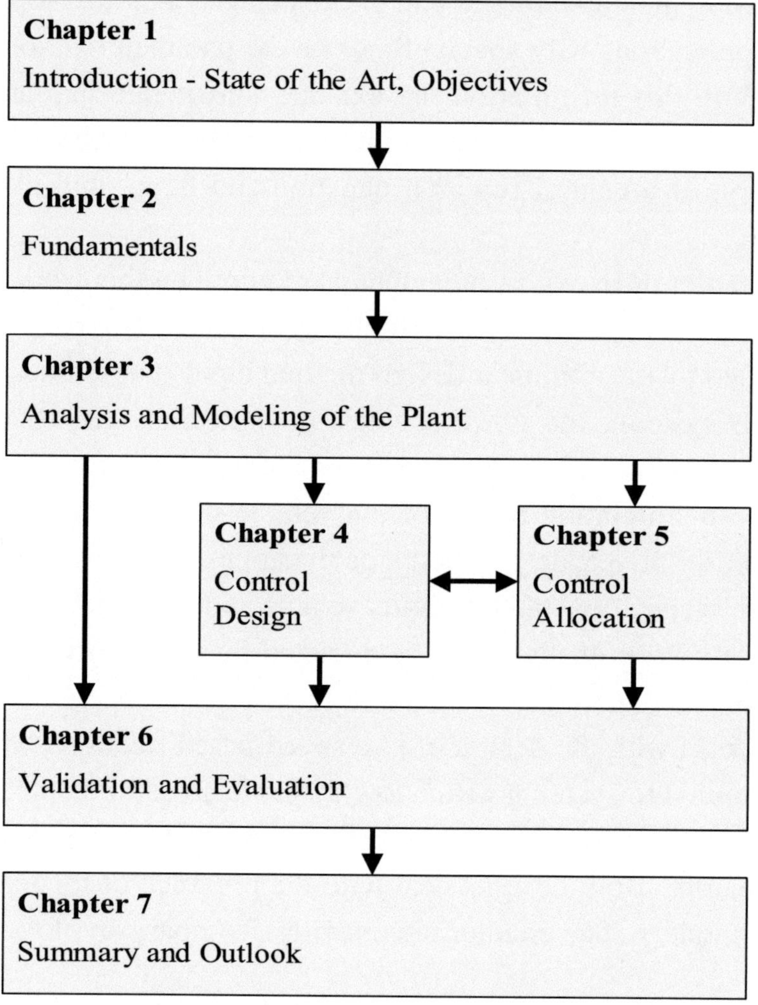

Figure 1.12.: Structure of the Work

2 Fundamentals

This chapter deals with preliminary considerations. First, the architecture of HEVs is described including the powertrain configuration and communication structure. This is followed by the fundamentals of control design and its performance parameters. The basic concept of the applied method input-output linearization is discussed. The chapter closes with presenting known methods for control allocation.

2.1. Hybrid Vehicles

Hybrid vehicles make use of at least two different primary energy sources. These are mostly gasoline stored in tanks and electric energy stored in batteries. These primary energy sources are then converted to mechanical energy by an ICE and EM respectively which then propels the vehicle. In general, both motors run simultaneously.

HEVs can be categorized by the degree of hybridization. It refers to the ratio of electric drive on the total traction system. With increasing maximum torque of the EM, its operating voltage rises as well. Increments in the automotive industry are 12 V, 48 V, 400 V and 800 V. The battery capacity also increases with the voltage level enabling vehicles to travel fully electric for an extended period of time. The highest category in this classification are Plug-In HEVs (PHEV) with HV batteries that can be charged from the grid. Another classification of HEVs is the topology of the powertrain [108].

Depending on the configuration and mechanical connection of these, different types of HEVs can be classified. Three groups are distinguished: **Series**, **Parallel** and **Mixed** hybrids. All groups further divide into subgroups, see figure 2.2. The schematics of each configuration will not be depicted. Details can be found in the works [56, 108, 107].

Series HEVs have no mechanical connection of the ICE to the wheels. The ICE propels an EM as generator that produces electric energy which is stored

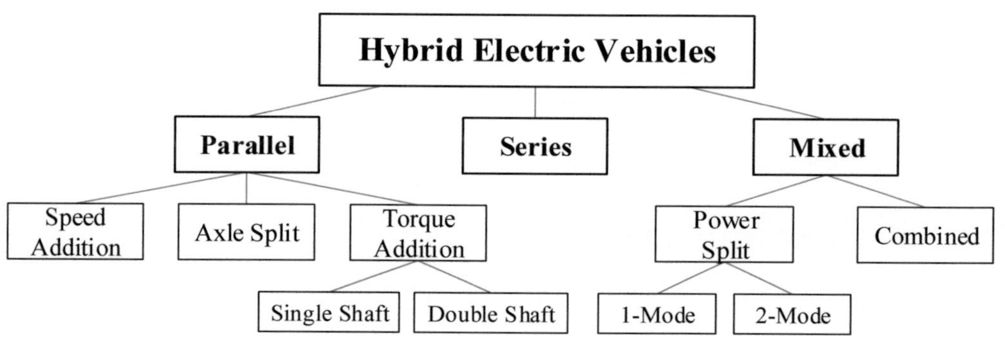

Figure 2.1.: Overview of HEV Powertrain Configurations - adapted from [11]

in a HV battery. An additional EM then converts this stored electric energy to mechanical which propels the vehicle. This configuration has the advantage that the ICE can be operated in efficient load points overcoming the losses due to several energy conversions. The number of EMs producing the tractive force can be one or up to the number of wheels of the vehicle.

Parallel HEVs are more diverse. Here ICE as well as EM have a mechanical connection to the wheels. Torque addition is achieved when an EM is added to a regular powertrain. It can be placed either directly on the crankshaft, between crankshaft and gearbox or between gearbox and wheels. Due to this parallel connection the torque output of the system is the sum of ICE and EM with respect to their individual transmission ratios. This group further divides into single-shaft and double-shaft concepts depending on the gearbox type. Only marginal differences in the controls result as the gear ratio between ICE and EM are either fixed or dependent on the selected gear. Axle split HEVs are propelled by the ICE on one axle exclusively whereas the other axle is driven by the EM. Speed addition is achieved with both motors and the drivetrain connected to a planetary gear. Here, the torque ratio is fixed limiting the degree of freedom of this concept.

The last category refers to mixed hybrids. Power split hybrids are similar to speed addition parallel HEVs but make use of an additional EM located on the cardan shaft adding a DOF for the operation strategy. A second planetary gear set is added for two-mode power split HEVs increasing the DOFs further. Combined HEVs are constructed to operate in both, series or parallel mode.

This work addresses hybrid traction control, using two actuators to fulfil a single control task. This requires a powertrain configuration with both motors mechanically connected to the same axle. Considering this the set of hybrid vehicles reduces to power split as well as parallel hybrids using torque and speed addition[1]. Due to the fixed torque ratio speed addition concepts will not be addressed as the DOF of interest is non-existent. The higher complexity of the powertrain system in power split hybrids leads to an increased cost [11]. Therefore, many manufacturers focus solely on parallel hybrids. Hence, this work will deal with this set of hybrid vehicles. In the following, torque addition parallel hybrids are referred to as parallel hybrids for simplicity reasons.

2.2. Vehicle Electronics

2.2.1. Bus Systems

State-of-the-art vehicles make use of various actuators throughout the vehicle which are controlled by individual ECUs. Communication with one another is enabled by bus systems. There are various topologies to connect ECUs by bus systems. Drivetrain ECUs and ECUs involved in driving dynamics are connected by a line topology, see figure 2.2. Several ECUs are attached to a single bus. To couple various bus systems, gateways are required due to the usage of different communication protocols and characteristica. Depending on the necessary data transfer rate different systems are used. Important bus systems in the automotive industry are the controller area network (CAN) for low and medium bandwidths of up to 1 Mbit/s, and FlexRay for high bandwidths of up to 10 MBit/s [106].

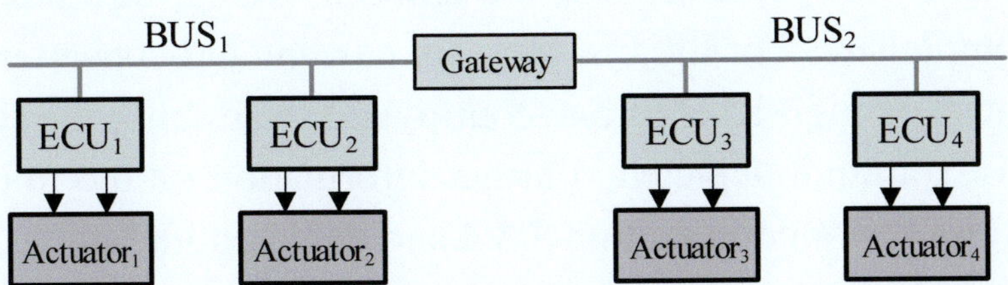

Figure 2.2.: Communication of ECUs with Bus Systems - adapted from [106]

2.2.2. Torque Control

To propel the vehicle, torque must be generated by the motors. The calculation of the desired torque depends on the accelerator pedal position and various

---

[1] Combined hybrids are not mentioned as they can operate as torque addition parallel hybrids. Therefore ideas for parallel torque addition hybrids also hold for combined hybrids

limitations. Figure 2.3 depicts the top-level torque control. These are level 1 functions which determine the functionality. Level 2 functions are diagnostic functions that monitor level 1 functions and can deactivate output stages of the ECU in case of failures. They are not displayed in figure 2.3. The main input to the system is the accelerator pedal position α_{pedal}. It is converted to a desired torque T_{pedal}. This torque is then filtered[2] for a smoother pedal feeling yielding the driver torque demand T_{driver}. Thereafter, torque demands from various requesters are coordinated to T_{coord}. This torque is then allocated to the individual low-level controls of the motors as desired torque T_d [1].

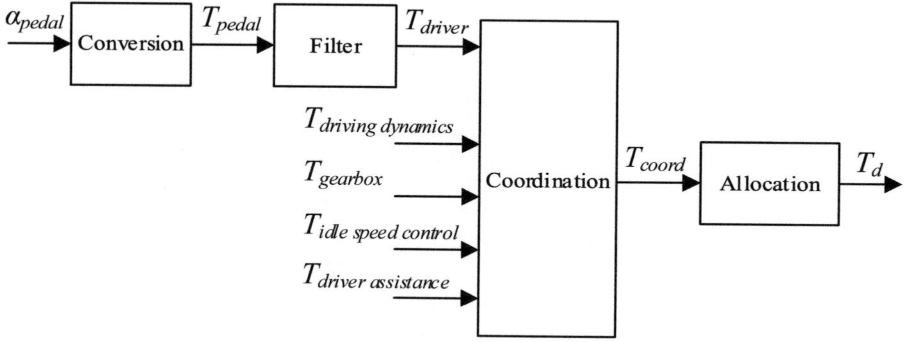

Figure 2.3.: Torque-Oriented Control - adapted from [106]

The desired torque is then communicated to low-level controls (LLCs) to regulate the demand. For the ICE these are ignition timing, throttle, intake valve timings, among others. For the EM they are current and converter control etc. Due to the focus on high-level torque control and to simplify the schematic the individual LLCs are not depicted. Further information on the functionality of the actuators can be found in section 3.3.4 and the cited literature.

2.3. Control Design

The standard control loop is depicted in figure 2.4. Here $P(s)$ is the to controlled plant and the $C(s)$ depicts the controller with s being the complex Laplace operator. The internal signals are defined as:

---

[2]Mostly by rate limitations depending on the absolute torque.

r reference signal   d load disturbance
e control error η process output
u control signal   n measurement noise
y measured signal   v plant input

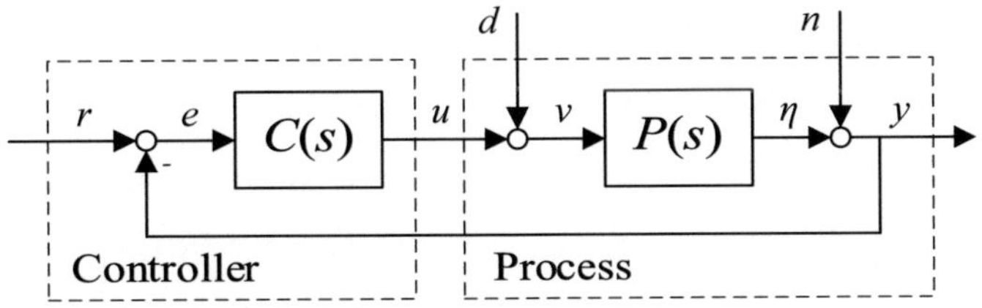

Figure 2.4.: Standard Control Loop - adapted from [9]

The linear plant $P(s)$ can be expressed in the state space form

$$\dot{x} = Ax + B(u + d)$$
$$y = Cx + D(u + d) + n, \tag{2.1}$$

where x is the state vector and A is defined as the system matrix. It describes the dynamic behavior of the plant. The input matrix is B and the output matrix is C. The feed-through (or feed-forward) matrix is described by D.

The standard control loop has several in and outputs. Depending on the resulting transfer functions the controller can be benchmarked. The transfer function matrix U describes the relationship between all inputs and outputs of the system. For the remainder of this section the complex Laplace operator s will be dropped.

$$\begin{bmatrix} y \\ \eta \\ v \\ u \\ e \end{bmatrix} = \underbrace{\begin{bmatrix} T & S_P & S \\ T & S_P & -T \\ S_C & S & -S_C \\ S_C & -T & -S_C \\ S & -S_P & -S \end{bmatrix}}_{U} \begin{bmatrix} r \\ d \\ n \end{bmatrix}. \tag{2.2}$$

The matrix U only consists of four different transfer functions which will be described in the following. They are often called **the gang of four** and define the characteristics of the closed-loop system.

$$S = \frac{1}{1 + CP} \qquad S_P = \frac{P}{1 + CP} \qquad T = \frac{CP}{1 + CP} \qquad S_C = \frac{C}{1 + CP}$$

The **sensitivity function** S describes the response from noise to the measured output and to the control error. It is also known as the disturbance transfer function. The **complementary sensitivity function** T describes the response from the reference signal to the control output. It is also known as the control transfer function. The **load sensitivity function** S_P describes the response from disturbances to the measured output and is also known as input sensitivity function. The **noise sensitivity function** S_C describes the response from the reference signal to the plant input and is also known as output sensitivity function [9].

All transfer functions have the same denominator, $1 + CP$, where

$$L = CP, \tag{2.3}$$

is considered as open-loop transfer function. Additionally,

$$S + T = 1, \tag{2.4}$$

which results in a conflict during control design. In the lower frequency spectrum with respect to the gain crossover frequency of the open-loop system ω_{gc} good disturbance attenuation is demanded:

$$|S(j\omega)| \ll 1 \quad \text{for} \quad \omega < \omega_{gc}. \tag{2.5}$$

For high frequencies, the influence of noise upon the control error becomes relevant. For this reason, it must be suppressed for frequencies higher than the crossover frequency:

$$|T(j\omega)| \ll 1 \quad \text{for} \quad \omega > \omega_{gc}. \tag{2.6}$$

2.3.1. Controllability and Observability

Fundamental properties that influence control design are controllability and observability of a system. Therefore, both definitions will be stated in the following.

Definition (Controllability). A linear system is controllable if for any $x_0, x_f \in \mathbb{R}$ there exists a $T > 0$ and $u : [0, T]$ such that the corresponding solution satisfies $x(0) = x_0$ and $x(T) = x_f$ [9].

Theorem 4.1 (Controllability rank condition). A system is controllable if the controllability matrix S_c has full rank

$$\text{rank}(S_c) = n. \tag{2.7}$$

The controllability matrix is defined as

$$S_c = [B, AB, A^2B, \cdots, A^{n-1}B]. \tag{2.8}$$

Definition (Observability). A linear system is observable if for any $T > 0$ it is possible to determine the state of the system $x(T)$ through measurements of $y(t)$ and $u(t)$ on the interval $[0, T]$ [9].

Theorem 4.2 (Observability rank condition). A system is observable if the observability matrix S_o has full rank

$$\text{rank}(S_o) = n. \tag{2.9}$$

The observability matrix is defined as

$$S_o = [C, CA, CA^2, \cdots, CA^{n-1}]. \tag{2.10}$$

2.3.2. Stability

Another fundamental criterion for dynamic systems is stability. The most prevalent are input to state stability and input output stability.

Definition (Lyapunov stability). The state of equilibrium $x_e = 0$ of the system (2.1) with $u(t) = d(t) = 0$ is stable, if for any $\varepsilon > 0$ there exists a number $\delta > 0$, so that for any arbitrary initial state, that satisfies the condition $\|x_0\| < \delta$ the eigen movement of the system (2.1) satisfies the condition $\|x(t)\| < \varepsilon$. The state of equilibrium is asymptotically stable, if it is stable and $\lim_{x \to \infty} \|x(t)\| = 0$ is satisfied [87].

Definition (Input output stability). A linear system (2.1) is input-output stable if for an initial state of zero $x_0 = 0$ and an arbitrary bounded input signal $|u(t)| < u_{max}$ the output signal also remains bounded $|y(t)| < y_{max}$ [87].

If a system is asymptotically stable it is also input output stable. Input output stability can be proven by examining the open loop transfer function.

Theorem 4.3 (Nyquist criterion). Let L be the loop transfer function for a negative feedback system (as shown in figure 2.4) and assume that L has no poles in the closed right half-plane (Re $s \geq 0$) except for single poles on the imaginary axis. Then the closed-loop system is stable if and only if the closed contour given by $\Omega = \{L(j\omega) : -\infty < \omega < \infty\} \in \mathbb{C}$ has no net encirclements of the critical point $s = -1$ [9].

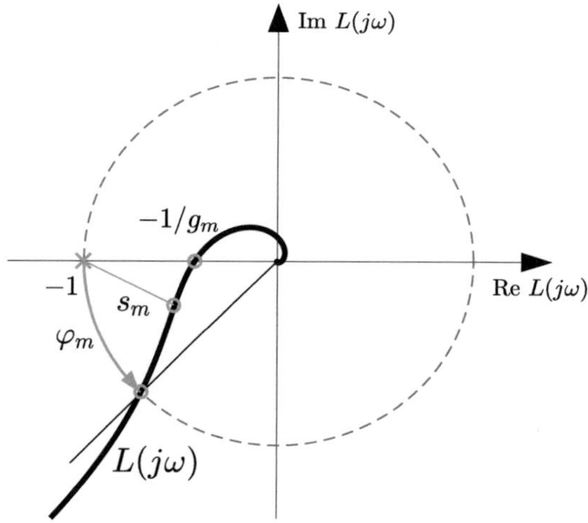

Figure 2.5.: Nyquist Plot - Phase, Gain and Stability Margin - adapted from [9]

For applications it is not enough to demand stability alone. There must also be margins of stability for robustness to perturbations. These can be derived directly from the Nyquist plot depicted in figure 2.5

The gain margin is the smallest amount that the open-loop gain can be increased before the system becomes unstable. It is defined as the inverse of the distance between intersection point of the locus curve with the real axis and the origin with ω_{pc} being the phase crossover frequency:

$$g_m = \frac{1}{|L(j\omega_{pc})|} \tag{2.11}$$

The phase margin is the amount of phase delay required to reach the stability limit. It is defined as the angle between the real axis and the line between origin and point on the locus curve crossing a circle around the origin with radius 1.

$$p_m = 180° - |\text{arg}L(j\omega_{gc})| \qquad (2.12)$$

with ω_{gc} being the gain crossover frequency. A relation of the demanded system damping and phase margin was given by Lunze [87]:

$$\zeta \approx \frac{p_m}{100°}. \qquad (2.13)$$

The stability margin is defined as the shortest distance between locus curve and the real axis at -1.

$$s_m = \min_{\omega} |1 + L(j\omega)| \qquad (2.14)$$

2.3.3. Loop Shaping

Designing T and S directly is difficult due to the nonlinear influence of the parameters in C on the closed-loop behavior. For this reason, SISO system controllers are often designed by specifications of the open-loop transfer function L. The advantage of this method is the linear dependency of the parameters in C on L. Additionally, the dynamic properties of the open-loop closely relate to the properties of the closed-loop transfer function. This method is called loop shaping. Figure 2.6 (a) shows a desired gain curve of a typical loop trans-

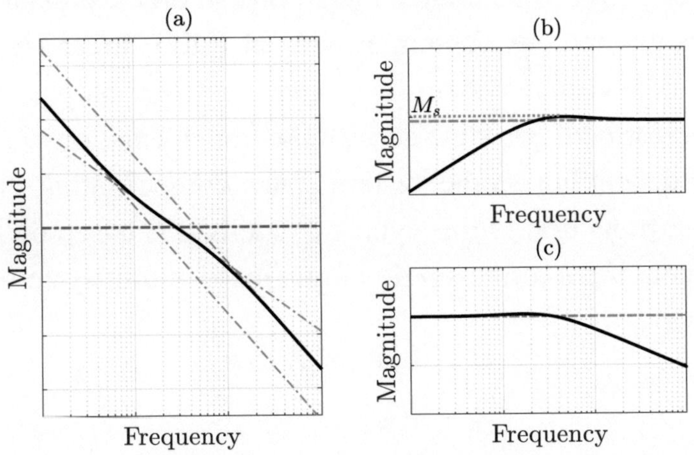

Figure 2.6.: Open-Loop Gain Curve (a), Sensitivity Function (b) and Complementary Sensitivity Function (c) - adapted from [9]

fer function. To ensure good reference tracking and load disturbance rejection, high magnitudes of the gain curve are desired for low frequencies, calling for a slope of $-40\,\frac{\text{dB}}{\text{dec}}$. The slope around the gain crossover frequency determine the robustness of the closed-loop system. It is desirable to choose a high gain crossover frequency for fast tracking. This though is limited due to robustness specifications of the stability margins. A desirable slope therefore is $-20\,\frac{\text{dB}}{\text{dec}}$. For high frequencies the slope is reduced to $-40\,\frac{\text{dB}}{\text{dec}}$ to prevent amplification of measurement noise [9, 29].

Further specifications can be derived from the sensitivity and complementary sensitivity functions in 2.6 (b) and (c) respectively. The disturbance attenuation is visualized by the sensitivity function. Its maximum value is an important variable giving the largest amplification of disturbances at a given frequency:

$$M_S = 1/s_m,\qquad(2.15)$$

which is the reciprocal value of the stability margin s_m defined in equation (2.14). Usually the maximum sensitivity is demanded [87]:

$$M_S \leq 2.\qquad(2.16)$$

2.3.4. Input-Output Linearization

The methods and definitions previously described only apply to linear systems. Nonlinear systems have to be treated differently in order to make them applicable to linear design methods. The main idea of IOL is to find a linear relationship between a new virtual system input and the actual output of the nonlinear plant. This is performed by a state coordinate transformation which generates a linearizing law depending on the system states compensating the nonlinearity. The following introduction of this synthesis method is based on [4, 63, 70].

First, a nonlinear input affine single input single output system of the form

$$\dot{x} = f(x) + g(x)\,u,$$
$$y = h(x)\qquad(2.17)$$

is considered where f, g and h are sufficiently smooth in the domain $D \subset R^n$. If the system is given in canonical form the linearizing law can be determined directly. If this is not the case then the output equation has to be transformed to its canonical form. This is done by differentiating the system output several times until the ρth derivative is a function of the nonlinear system input u:

$$y^{(\rho)} = f(u)\,. \tag{2.18}$$

Note that $\rho \leq n$ where n is the order of the system defined in equation (2.17). Setting $y^{(\rho)}$ to

$$y^{(\rho)} = -a_{\rho-1}y^{(\rho-1)} - \cdots - a_1\dot{y} - a_0 y + Vw\,. \tag{2.19}$$

an arbitrary dynamic behavior can be defined for the resulting linear system by choosing the coefficients a_i and V freely. With this transformation the nonlinear system input-output behavior is linearized between the new virtual input w to output y. Therefore, the name *exact input-output linearization* arises. Solving equation (2.19) for u, present in the derivative $y^{(\rho)}$, the following depiction in figure 2.7 becomes apparent. The derivation of IOL is found in appendix A.1.

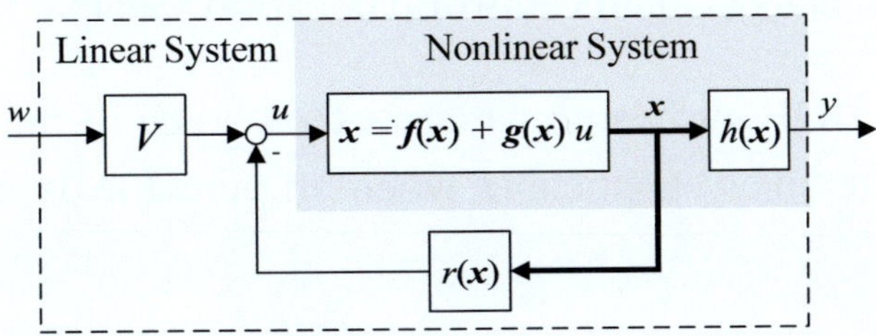

Figure 2.7.: Structure of the Resulting Linear System - adapted from [4]

2.4. Control Allocation

Overviews on control allocation methods were summarized by Härkegard [53] and Johansen [67]. They range from simple solutions such as direct control allocation distributing the control effort among the actuators by geometric reasoning [31, 32]. More sophisticated solutions will be addressed in the following.

2.4.1. Daisy Chain

The daisy chain control allocation allocates all the control effort v to the primary actuator considering its position and rate limits. Until these are reached only the primary actuator is addressed. When its limits are reached the difference in control effort is applied to the second actuator. If this, again, reaches its limits, then the next actuator is stressed with the remaining control effort. This goes on until all actuators are acting at their limits. The idea was proposed by Bordington [15]. A mathematical description as well as a Lyapunov stability analysis of the method can be found in [18]. A graphic illustration of the method is depicted in figure 2.8. It shows the daisy chain method for a system with two actuators. This method is a common choice of control allocation as it is simple.

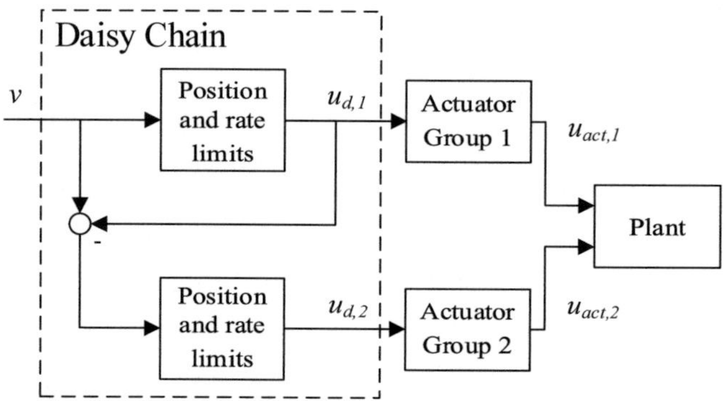

Figure 2.8.: Daisy Chain Control Allocation - adapted from [51]

There is only one DOF for design which is the choice of the actuator priority. This method is effectively applicable when all actuator dynamics are similar [51].

2.4.2. Optimization-Based Methods

Optimization based methods determine the control inputs u by solving a two-step optimization problem. This problem formulation is given by

$$
\begin{aligned}
u &= \arg \min_{u \in \Omega} \left\| W_u(u - u_d) \right\|_p, \\
\Omega &= \arg \min_{u_{min} \leq u \leq u_{max}} \left\| W_v(Bu - v) \right\|_p,
\end{aligned}
\tag{2.20}
$$

where u_d is the desired control input. First, Ω, the set of feasible control inputs, is determined by minimizing $Bu - v$, see equation (1.2). Given Ω, the control input u is chosen by minimizing $u - u_d$. Both terms are weighted by the matrices W_u and W_v which can be used for prioritization of either part.

The problem description (2.20) is subject to a still to be selected norm of the order p. The l_1 norm will lead to a linear problem formulation which is usually solved with the simplex method [85]. The l_2 norm leads to a quadratic problem, which is the most frequently used norm. To find a solution for this computationally intensive problem several numerical methods exist. They are described thoroughly in [51].

Independent from norm and numerical method the previously described control allocation method will be referred to as static control allocation. This method has no dependency on previous control inputs and treats every actuator independent from its dynamic properties. This leads to problems in systems composed of actuators with highly different actuator responses.

2.4.2.1. Dynamic Control Allocation

DCA is an extension to the static allocation approach and was introduced by Härkegard [51, 52, 53]. It explicitly takes the dynamics of the control signal into account. Using the l_2 norm the problem formulation is given by

$$u(t) = \arg\min_{u \in \Omega} \|W_1(u(t) - u_s(t))\|_2^2 + \|W_2(u(t) - u(t-T))\|_2^2,$$

$$\Omega = \arg\min_{u_{min} \leq u \leq u_{max}} \|W_v(Bu(t) - v(t))\|_2 \tag{2.21}$$

where u is the true control input, u_s is the desired steady state control input, v is the virtual control input and $\|W_2(u(t) - u(t-T))\|_2^2$ is the previous control input. Again, B is the control effectiveness matrix and W_1, W_2 and W_v are square matrices of proper dimensions for weighting certain terms of the quality function.

This formulation extends the static control allocation in (2.20) by the term $u(t) - u(t-T)$ incorporating the dynamics of the control input into the optimization problem. With this extension the desired actuator rates can be penalized. This method can be extended via a model predictive approach in order to satisfy underlying constraints to a greater detail [113].

The Non-Saturated Case

If no actuator saturation occurs the problem (2.21) can be rewritten to

$$\min_{u(t)} \|W_1(u(t) - u_s(t))\|_2^2 + \|W_2(u(t) - u(t-T))\|_2^2,$$

$$\text{subject to } Bu(t) = v(t) \tag{2.22}$$

by disregarding actuator constraints. Then, the explicit solution of (2.22) is given by

$$u(t) = Eu_s(t) + Fu(t-T) + Gv(t) \tag{2.23}$$

where

$$E = (I - GB)W^{-2}W_1^2$$
$$F = (I - GB)W^{-2}W_2^2 \tag{2.24}$$
$$G = W^{-1}(BW)^{-1})^\dagger.$$

The operator † denotes the pseudoinverse which is defined as

$$A^\dagger = A^T(AA^T)^{-1}$$

and the matrix W is given by

$$W = \sqrt{W_1^2 + W_2^2}.$$

Equation (2.23) shows that the solution of the optimal control allocation problem (2.22) is given by sets of linear filters. This though only applies if no actuator saturation is present. The expression of the linear filter in (2.23) can now be rewritten into discrete state space form. Then, the set of filters are given by

$$x(t+1) = Fx(t) + F(Eu_s(t) + Gv(t))$$
$$u(t) = x(t) + Eu_s(t) + Gv(t) \tag{2.25}$$

The states in x are now dependent on the previous states as well as the inputs u_s and v. Note that the system output is defined as u, the actual control input. The explicit solution is depicted in figure 2.9. The distribution of control effort among the actuators is defined by the properties of a set of linear filters.

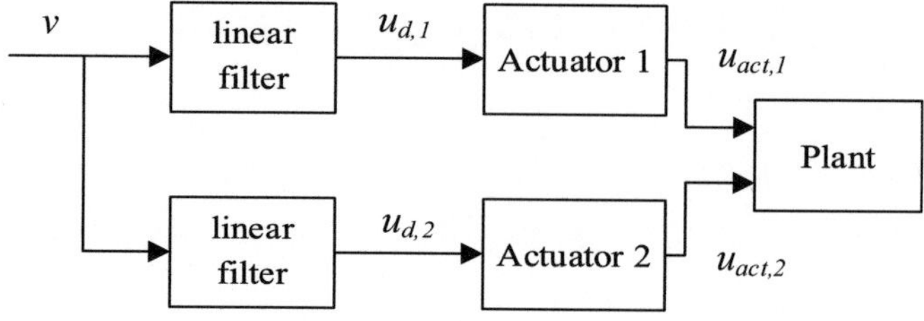

Figure 2.9.: Dynamic Control Allocation - Explicit Solution

3 Analysis and Modeling of the Plant

This chapter deals with a model of the plant dynamics. Initially, requirements for simulation models will be discussed. Then, the test vehicle and its powertrain configuration will be presented. Based on this, a model of the driving dynamics, drivetrain, actuators, and ECU network will be derived. The model will be derived from physical considerations. Unknown parameters will be determined through identification measurements. The verified model is the basis for control and allocation design in chapters 4 and 5, respectively. The chapter closes with an open-loop verification of the model.

3.1. Model Requirements

Before developing a model for this work, modeling requirements need to be addressed. No absolute guidelines on how to choose the correct model exist. "The choice of the best model is regarded more of an art than a science" [17]. For this choice Brooks et al. propose several aspects that need to be considered. These can be categorized by four major elements:

1. Results

2. Future Use of the Model

3. Verification and Validation

4. Resources Required

The element **Results** considers the extent to which the model output describes the behavior of interest. Also, the accuracy and the ease with which the model and its results can be interpreted belong to this element. The element **Future Use of the Model** refers to the portability of the model and the integrability with other models. **Verification and Validation** describes the probability of the model containing errors. Also, the accuracy of the model matching measurement data is addressed here. The last aspect to be considered when obtaining a model is **Resources Required**. It addresses time and cost to build, maintain the model but also to analyze its results. Additionally, hardware requirements must be taken into account. In the following, all four elements will be examined to generate a model that meets these proposed elements.

After the modeling guidelines were described, now the test vehicle will be presented. From this, a mathematical model of the system will be derived. The focus here lies on the depiction of the fundamental dynamics occurring during marginally stable maneuvers. Although developing a model for simulation purposes it will be modeled control oriented. This is motivated by reasons of simplicity to keep the effort at a minimum for follow up studies. Also, the developed controller in this work will be tested and validated on an industrial vehicle. This makes the derived model a first verification step of the controller development process. The main use of the model is to verify the controller and to parameterize the control allocation designs.

3.2. Test Vehicle

The investigated vehicle is a BMW 740 Le, a plug-in HEV. It has rear wheel drive and belongs to the category of single-shaft torque addition parallel HEVs.

The powertrain configuration and mechanical connection of the components are depicted in figure 3.1. As powertrain the complete set of driveshafts from engine to wheels also including the motors is defined. The drivetrain on the other hand only incorporates the shafts and gearboxes. The focus of the powertrain description lies on a system-level viewpoint of the main driveline components.

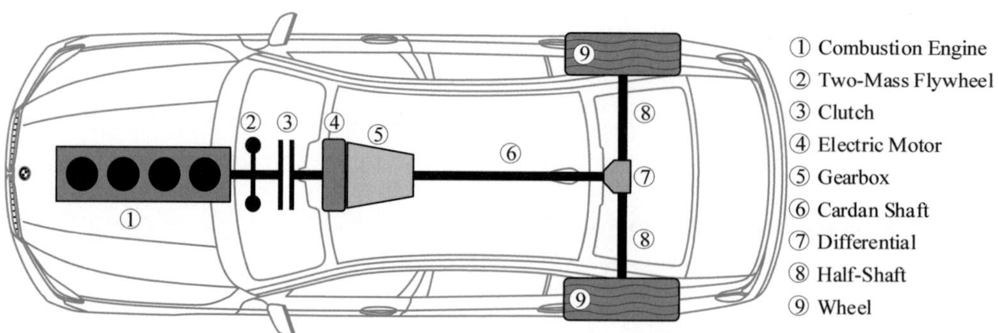

Figure 3.1.: Powertrain Configuration of a P2 HEV - adapted from [56]

The vehicle has two drives, a turbocharged ICE and a permanent magnet synchronous EM. The EM rotor is connected to the input shaft of the gearbox. The crankshaft of the ICE is connected to a two-mass flywheel. This in turn is connected to the EM rotor through a clutch. The gearbox output connects to the cardan shaft which moves the differential. The wheels are driven by half-shafts which connect to the outputs of the differential. The technical details of this powertrain system can be found in table 3.1. The tire specifications are $225/60R1799Y$. The vehicle was equipped with these for all experiments: identification, verification, and validation tests.

Table 3.1.: Technical Data of Considered Hybrid Vehicle

Parameter	Value	Unit
Internal Combustion Engine		
Engine Displacement	1998	cm^3
Number of Cylinders	4	-
Peak Torque	400	Nm
Peak Power	190	kW
Electric System		
Peak Torque	250	Nm
Peak Power	83	kW
Energy Content of HV Battery	11	kWh
Combined System		
Peak Torque	500	Nm
Peak Power	240	kW

Since the vehicle is a plug-in HEV it supports pure electric drive due to its high battery energy content as well as torque output of the EM. Figure 3.2 shows the maximum torque and power of ICE and EM. The ICE has twice the power and torque of the EM. The full torque though can only be provided from $1500\frac{1}{\text{min}}$ onwards whereas the EM already provides full torque from low rotational speeds.

3.3. Vehicle Model and Identification of Parameters

This section explores the development of a suitable model for investigating TC of a parallel HEV. First, the background of driving dynamics is described. This is followed by the depiction of the drivetrain dynamics. Then, the actuator dy-

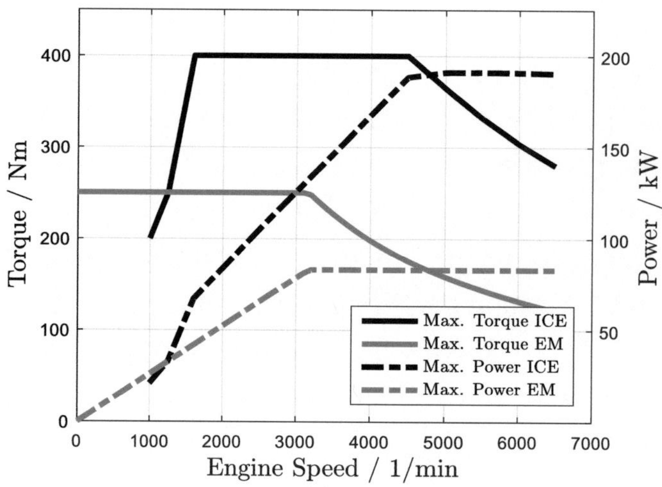

Figure 3.2.: Maximum Torque and Power of ICE and EM

namics will be dealt with. The section closes with an identification of communication delays between ECUs. The presented simulation model will be derived analytically based on the governing physical relationships and their dynamic equations. It will be used as basis for controller and allocation algorithm design in chapter 4 and 5. Parameters that are unknown will be identified. A verification through vehicle tests of various model parts will be provided. The identification process was performed partly as proposed in [140] and [142]. This chapter focuses on the dynamic depiction of the plant. Model parameters and identification results are listed in appendix A.3.

The overview of the complete vehicle model is depicted in figure 3.3. The model has two main inputs, the position of the accelerator pedal α_{pedal} and the maximum friction coefficient of tire road adhesion μ_{max}. Initial values such as gear, velocity, and rotational speeds are also taken into account. The model is divided into four parts. The coordination and control is performed by the control network[1]. It provides the desired torque T_d to the second part of the model, the actuators. These transform the desired signals into actual values with their individual dynamic behavior. The block actuators therefore represents the LLC as well as the mechanical setup. The actual torque T_{act} then is input to the third part of the model, the drivetrain. It reflects the rotational degree of freedom by the connected shafts and gearboxes and outputs the wheel speed ω_{whl} and the controlled variable y. The wheel speed and maximum friction coefficient are input to the fourth part of the model, the driving dynamics. It outputs the

---

[1]Network of ECUs that communicate by bus

longitudinal tire force F_x which acts upon the drivetrain and the driving state which is measured by the ECUs within the control network.

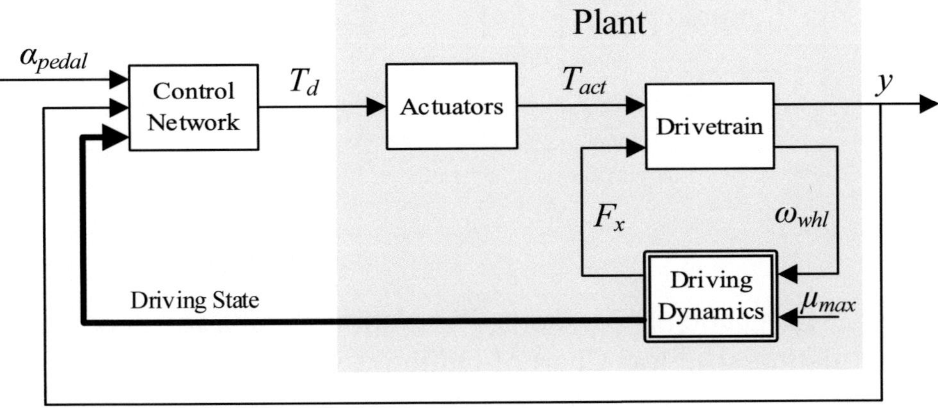

Figure 3.3.: Overview of Vehicle Model

3.3.1. Identification Procedure

In order to fit unknown model parameters to measurements of the modeled process, an identification is made. Process identification is described by Isermann [61] and was applied in [142]. First, measurements of the to be identified real system are made. Inputs $u_{measure}$ and outputs $y_{measure}$ must be tracked. Then, the to be identified model will be excited with the measured input in a simulation. The model output y_{model} will be compared to the system output and the error e determined, compare figure 3.4 (a). The root mean squared error (RMSE) is used in this work.

The identification process is an optimization problem. The flow chart for this is depicted in 3.4 (b). For the first simulation, initial parameters will be used. These will be updated after every simulation as long as the error or quality function in general is greater than a given threshold. The simplex method introduced by Nelder and Mead, a deterministic numeric algorithm is chosen to perform this optimization process [96]. This method produces a sequence of simplexes in the search area in a way that the error or quality function decreases on its edges. Although it may not find the exact minimum of the quality function, as opposed to gradient based approaches, this method is widely used in practice [99]. This is due to its robustness and fast convergence to minima when only using few search variables [115, 76]. Since only few variables need

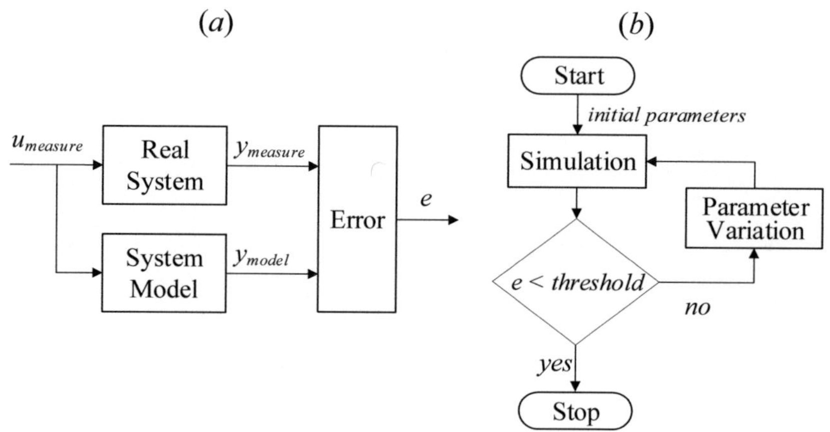

Figure 3.4.: Identification Process, (a) Illustration of Measurement Based Simulation, (b) Flow Chart of Optimization

to be optimized at once, and the start conditions are known well, this method is appropriate for the given problems.

3.3.2. Driving Dynamics

The driving dynamics will be depicted by a quarter car model. The effects of driving dynamics are based on forces and torques acting on the vehicle body defining the dynamic behavior of the system. These are depicted in figure 3.5. Also, lengths to define the center of gravity (COG) are shown. The longitudinal tire force accelerating the vehicle is described by F_x. Air and rolling resistance are F_a and F_r respectively. Torque provided from the drivetrain acting on the wheel is defined by T_{dt}. The mass of the vehicle is m and g represents the gravitational acceleration. The wheel radius is r_{whl} and the lengths l_f and l_r show the distance from the COG to the front and rear axle, respectively. The height of the COG with respect to the axles is given by l_h.

The central variable defining the longitudinal tire force is the wheel slip λ which was defined in equation 1.1 and is responsible for the tractive force accelerating the vehicle. The tractive force is represented by a function of the vertical force $F_z(a_x)$ which is a function of the longitudinal acceleration and the adhesion coefficient μ. It is a function of the wheel slip and given by

$$F_x = F_z(a_x)\mu_x(\lambda_x).$$

(3.1)

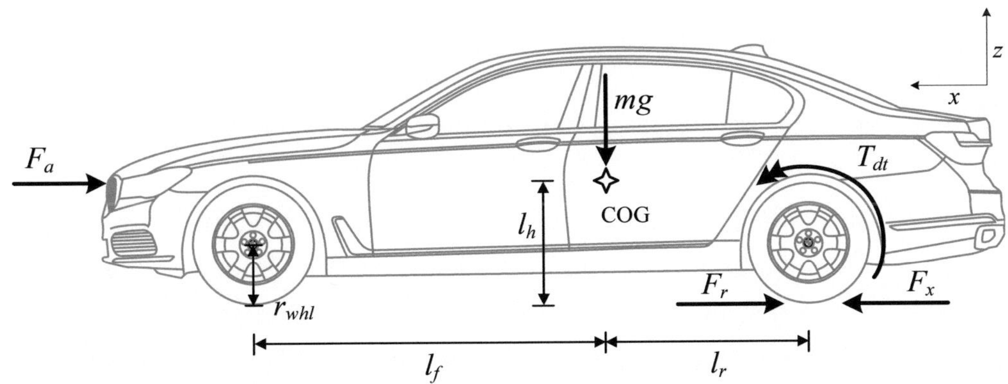

Figure 3.5.: Forces and Torques Acting on and Dimensions of the Test Vehicle

The characteristic of $\mu_x(\lambda_x)$ strongly depends on the tire-road conditions. The progression of the curve though always remains similar, see figure 1.1. In the beginning there is a high pseudolinear gradient which then flattens forming a distinct maximum. After this, the slope has a negative gradient until it reaches its definition limits at $\lambda = 1$.

As seen in figure 1.1 the adhesion coefficient has a nonlinear dependency on the wheel slip. Since there is no physical relationship, functional expressions were determined empirically through various static measurements varying the driving force and the speed. The Magic Formula was introduced by Hans B. Pacejka et al. [10, 98] and is a widely used representation [137, 23, 36, 117]. Therefore, it is chosen as a tire model in this work. The mathematical expression is given by equation 3.2.

$$\mu_x(\lambda_x) = \mu_{max} \sin(C_{pac} \arctan(B_{pac}\lambda_x - E_{pac}(B_{pac}\lambda_x - \arctan(B_{pac}\lambda_x)))) .$$

$$(3.2)$$

The maximum friction coefficient of the tire-road adhesion is described by μ_{max}. Parameters B_{pac}, C_{pac} and E_{pac} influence the progression and shape of the function. The parameters for the simulation model were identified in an open-loop experimental study encountering a large spectrum of slip values.

The vertical force in equation 3.1 is subject to the vehicle acceleration due to the dynamic axle load distribution. If the COG of the vehicle is not aligned with front and rear axle then a force acted upon this mass exerts a pitching force. Therefore, the vertical force will be represented as a function of the longitudinal acceleration. The static part is defined by the mass of the vehicle and the location of the COG with respect to the axles. The dynamic part for the rear axle uses the vehicles acceleration a_x as well as the distance l_h.

$$F_z(a_x) = F_{z,static} + F_{z,dynamic}(a_x) = m_{veh}g\frac{l_f}{l_f + l_r} + m_{veh}\frac{l_h}{l_f + l_r}a_x. \qquad (3.3)$$

Air and rolling resistance are described in equation 3.4 and 3.5 respectively. The air resistance is a quadratic function of the vehicle velocity given by $v_x|v_x|$ to consider driving forwards and backwards. It also makes use of the front contact area A_{veh} of the vehicle, the drag coefficient c_x and the air density ρ. The rolling resistance F_r is a constant value for $v_x \neq 0$ and determined by the resistance coefficient e_r as well as the mass of the vehicle m_{veh} and the gravitational acceleration g.

$$F_a(v_x) = \frac{1}{2}\rho c_x A_{veh} v_x |v_x|, \qquad (3.4)$$

$$F_r(v_x) = e_r m_{veh} g \, \text{sign}(v_x). \qquad (3.5)$$

The balance of forces for the center of mass is described in equation 3.6. The driving force F_x accelerates the mass of the vehicle m_{veh} whereas air F_a and rolling resistance F_r act against it.

$$a_x = \frac{1}{m_{veh}}(F_x - F_a - F_r). \qquad (3.6)$$

The longitudinal torque acting on the wheel caused by driving dynamics is T_{whl}. It is the difference of driving force and rolling resistance multiplied by the wheel radius[2]:

$$T_{whl} = r_{whl}(F_x - F_r). \qquad (3.7)$$

Next to a quarter car more detailed models exist also describing cornering or vertical dynamics. These are single track or double track model which distinguish between forces and torques acting on individual wheels. Since this work focusses on the control of the over-actuated drivetrain and the inner loop of the speed-based traction controller, they will not be described here. Further information on these is found in [62, 98, 55].

---

[2]The air resistance only acts on the vehicle body

3.3.3. Drivetrain Dynamics

The drivetrain is the group of components that provide power to the driven wheels. It consists of connected shafts and gearboxes but excludes motors. As pointed out in the state of the art 1.2.1 most works on traction control consider a rigid drivetrain. Pinto et al. [103] describes works dealing with oscillation damping which calls for modelling a flexible driveshaft with an elasticity adding further DOFs to the model. Depending on the use of the model a two-mass oscillator or higher order systems are considered [83, 101]. In order to choose a suitable modeling complexity, the drivetrain dynamics will be analyzed. Starting point is a high degree elastic drivetrain model which was derived from component specifications. Further information and details on automotive vehicle drivetrains are described in [116, 3].

The drivetrain model used for analysis depicts five inertias and four drivetrain stiffnesses and dampers see figure 3.6. These are the two-mass flywheel, cardan shaft, half-shafts and torsion of the tire beld. The parameters are derived from the component design specifications. The parameterization is found in appendix A.3.

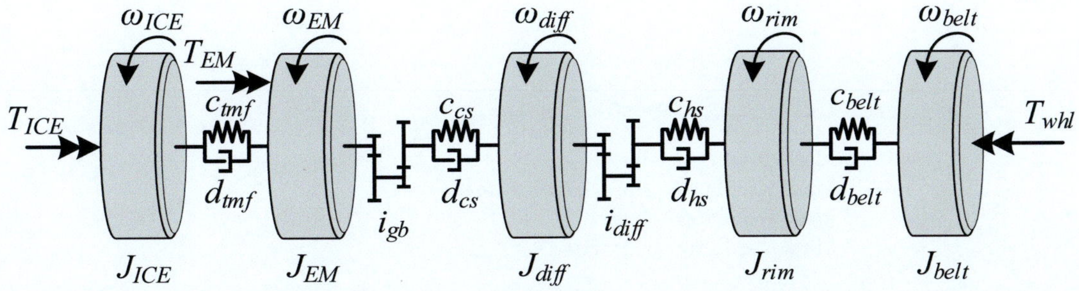

Figure 3.6.: Drivetrain Model - Five-Mass Oscillator

The drivetrain dynamics are highly nonlinear due to the tire-road coupling. For the analysis of the system therefore a linearization must be performed. To identify which linearization points are of interest known research will be examined. Bottiglione et al. present Bode plots of elastic drivetrain dynamics for a linearization around different slip values [16]. Their work shows a sharp increase in the eigenfrequency of the powertrain when the slip is increased to the region where traction reaches its maximum. For low slip values the eigenfrequencies of the drivetrain is also low. As soon as the critical slip value is reached, defining the peak of the μ-λ curve, the eigenfrequencies shift abruptly to higher values, compare figure 3.7. After this point no significant change in the eigenfrequency and amplitude is observable. This calls for a distinction between the drivetrain dynamics in the pseudo linear stable part and the nonlinear unstable part. A similar study is performed by Yeap et al. but with a different scope, characterizing stable and unstable oscillations [136]. Also, a high dependency of the actual slip on the drivetrain oscillatory behavior was shown with abrupt changes when the critical slip value is reached.

Regarding the research results discussed above, for this work two linearized models of the drivetrain will be analyzed. First there is the stable pseudo linear part and second the unstable nonlinear part. For a better differentiation, the linearized models of the pseudo linear part of the μ-λ curve will be named *linear model* and the linearized models of the nonlinear part *linearized nonlinear model*.

The Taylor Series Approximation around an operating point is used [4]. The linearization will be performed for the slip values $\lambda_{lin} = [0.01, \mu_{max}]$. This is sufficient to linearize the longitudinal tire force. Additionally, a fixed vehicle speed is assumed for the linearization of the slip definition $v_{lin} = 10$ m/s. This is also used to determine the air and rolling resistances in the chosen operating point.

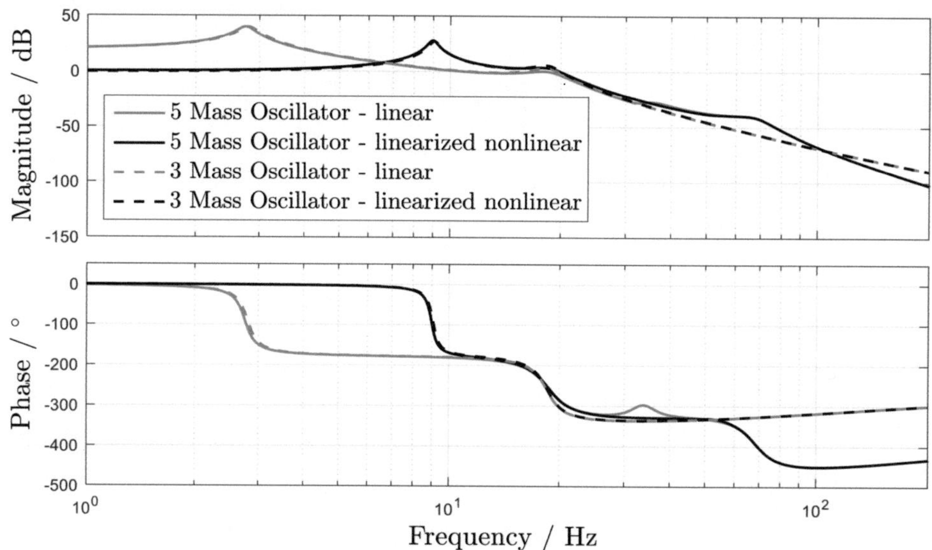

Figure 3.7.: Bode Plot of Drivetrain Dynamics from ICE Torque to Wheel Speed - linear / linearized nonlinear

Figure 3.7 shows the Bode plots of two linearized versions of the five-mass oscillator drivetrain model as well as the three-mass oscillator of reduced order from engine torque input to half-shaft torque output. The derivation of the three-mass oscillator is discussed later in this section. The initial amplitude for linear and linearized nonlinear model are different due to the tire force influence. The distinct peaks in the magnitude trace back to the elastically modeled drivetrain components. As Bottiglione already showed, a significant shift in the eigenfrequency of the half-shaft from linear to linearized nonlinear model can be observed. A shift in the eigenfrequency from 2.5 Hz to 8.5 Hz is observable. Also, the tire belt shows a dependency on the operating points. The figure also shows good conformance of the reduced three-mass oscillator model compared to the five-mass oscillator model for frequencies up to 30 Hz.

In tables 3.2 and 3.3 eigenvalues and eigenfrequencies of the linear and linearized nonlinear model are shown. Additionally, the dominance of each value is depicted. The dominance was introduced by Litz as a measure of the significance of a DOF [81]. For the analyzed drivetrain two eigenfrequencies are dominant for both models. These can be traced back to the elastic half-shaft and the two-mass flywheel and have eigenfrequencies at the lower end. The two higher eigenfrequencies have a low dominance.

Table 3.2.: Eigenvalues and Eigenfrequencies of Considered Five-Mass Oscillator with Dominance for Linear Plant

Eigenvalue	Eigenfrequency	Dominance
$\lambda_{0,1} = 0 \pm 0j$	$f_0 = 0$ Hz	∞
$\lambda_{2,3} = -0.84 \pm 17.28j$	$f_1 = 2.76$ Hz	0.5873
$\lambda_{4,5} = -13.79 \pm 116.38j$	$f_2 = 18.65$ Hz	0.1849
$\lambda_{6,7} = -20.26 \pm 216.92j$	$f_3 = 34.67$ Hz	0.0067
$\lambda_{8,9} = -50.90 \pm 426.20j$	$f_4 = 68.31$ Hz	0.0005

The model requirements in general favor a low DOF model due to handling and validation. Since two eigenfrequencies have a low dominance they will be dropped by performing a model reduction. Another reason to drop the depiction of these components is the human perception threshold on vibrations. The human perception intensity on vibrations though drops significantly for frequencies higher than 10 - 15 Hz, see figure 1.8. Further information on the human vibration perception can be found in the following literature [91, 131]. The model reduction will be performed by the method introduced by Laschet [77]. It is described in appendix A.2.

Table 3.3.: Eigenvalues and Eigenfrequencies of Considered Five-Mass Oscillator with Dominance for Linearized Nonlinear Plant

Eigenvalue	Eigenfrequency	Dominance
$\lambda_{0,1} = 0 \pm 0j$	$f_0 = 0\,\text{Hz}$	∞
$\lambda_{2,3} = -1.49 \pm 56.41j$	$f_1 = 8.98\,\text{Hz}$	10.2132
$\lambda_{4,5} = -13.79 \pm 116.74j$	$f_2 = 18.70\,\text{Hz}$	3.0975
$\lambda_{6,7} = -31.39 \pm 333.55j$	$f_3 = 53.32\,\text{Hz}$	0.1650
$\lambda_{8,9} = -51.01 \pm 426.38j$	$f_4 = 68.34\,\text{Hz}$	0.0152

For this work all drivetrain parameters were known a priori due to the availability of dimensioning information in the vehicle development process. If they are unknown, a verification process can be performed. A detailed identification process for the elastic vehicle drivetrain including the driving dynamics model was introduced in [140]. It was proposed to especially identify inertias and stiffnesses of the half-shafts as they are responsible for the torsional dynamics.

Concluding the considerations of this section the incorporation of half-shaft and two-mass flywheel stiffness is a well-balanced tradeoff between modeling accuracy and modeling effort especially for P2 Hybrid vehicles. The schematic of the model is depicted in figure 3.8.

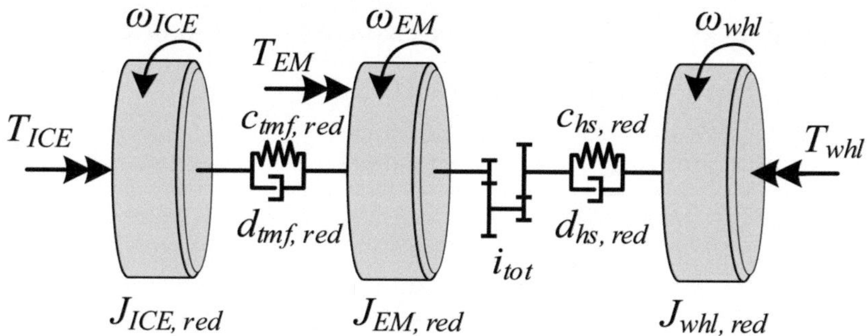

Figure 3.8.: Drivetrain Model - Three-Mass Oscillator

3.3.4. Actuator Dynamics

A HEV makes use of two actuators, an ICE and an EM. Therefore, a general control oriented model will be proposed for each actuator. Unknown abstract parameters will be identified followed by a verification of the dynamics. The models are based on publications [142] and [140].

3.3.4.1. Internal Combustion Engine

There are various levels of detail to model the torque generation of an ICE. Control oriented models make use of mean value models [47]. For these models, the assumption holds that the ICE produces torque continuously. The dynamics are modeled for the produced torque over the average of 720 degrees of crankshaft angle. Since this work explicitly focusses on actuator dynamics a more detailed approach is chosen combining a mean value engine model with a zero-dimensional thermodynamic model. This depicts the physical behavior to a higher degree of accuracy. It is derived from and verified with a gasoline engine but can also be applied to diesel engines as the underlying principles remain the same.

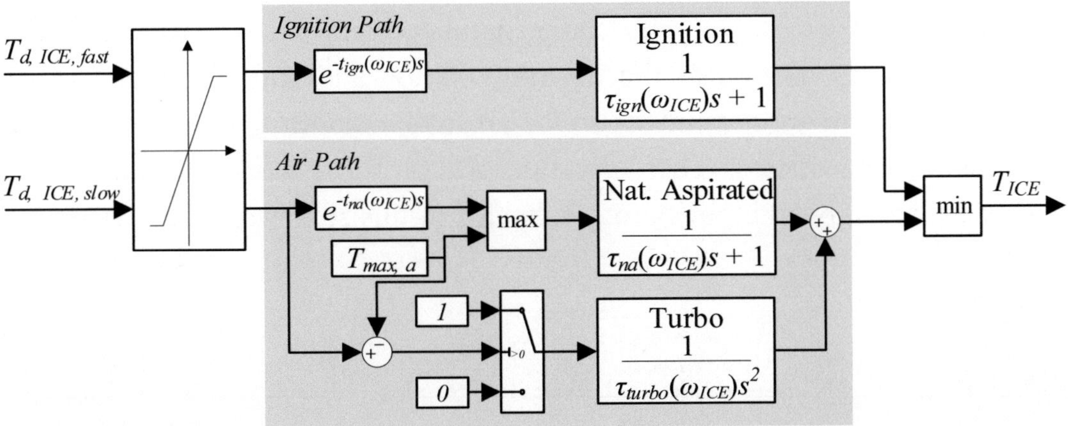

Figure 3.9.: Internal Combustion Engine Model - adapted from [142]

Before the developed model will be depicted, first the fundamental functionality of a gasoline engine will be described: air is pumped into a cylinder and gasoline is injected into the chamber which produces a flammable mixture. Then, an ignition spark ignites the mixture which causes it to convert the chemically stored energy into thermic energy which results in an expansion of the mixture. This produces a force acting upon the cylinder head generating a torque upon the crankshaft. Further details on the functionality of internal combustion engines can be found in [89, 12].

State of the art engines possess various actuators increasing the power and dynamics of the entire system. These are direct injection, full variable valve control, turbocharging and variation of ignition timing. An overall scheme of a model to describe this behavior is depicted in figure 3.9. Inputs are two desired torques $T_{d,ICE,fast}$ and $T_{d,ICE,slow}$. These are limited to the maximum producible torque depending on the rotational speed of the engine. The desired torques then excite the dynamics of the system. These can be categorized into fast **ignition path** and slow **air path** dynamics. The latter can be further divided into the **naturally aspirated** dynamics **turbo** dynamics. The turbo dynamics are only excited when the desired torque exceeds the limits of natural aspiration. The torque produced by the several subsystems limit each other. This results in only the minimum torque as the output. All time constants and time delays are a function of the engine speed. In the following the model will be described in detail and verification measurements will be provided.

Naturally Aspirated

It is important to notice that the torque that can be produced is limited to the amount of air in the cylinder. This can be regulated rather fast up to atmospheric pressure through variation of the throttle position and variable intake valves. The dynamics here are dependent on the air intake manifold dimensioning and its resonance frequencies. The derivation of a physical model does not justify the cause since only an input-output behavior of this subsystem is relevant.

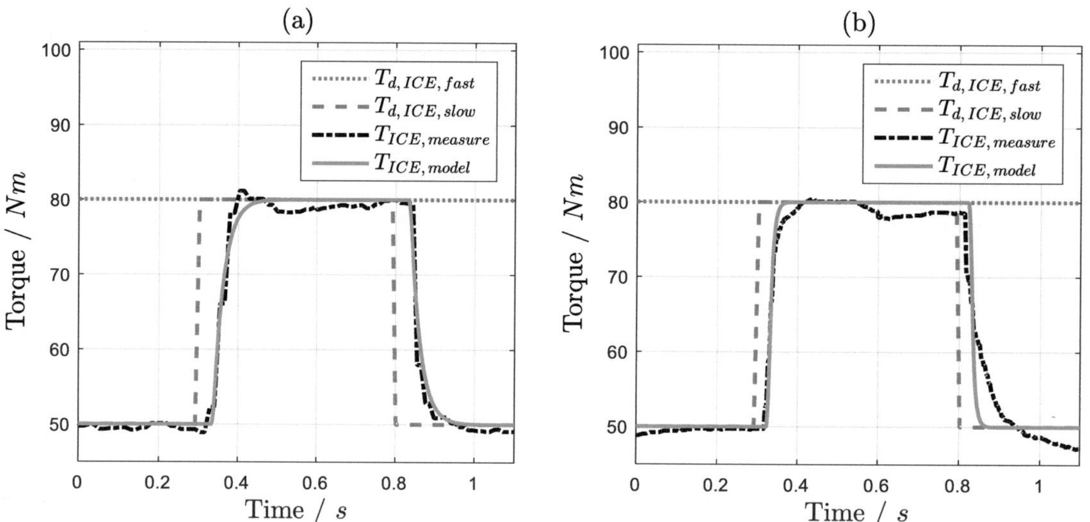

Figure 3.10.: Verification of Naturally Aspirated Air Path Model - Step Responses at 1500 rpm (a) and 5000 rpm (b) of Measurement Data and Model

The dynamics can be roughly modeled as a speed dependent first order low pass with a time delay [47]. The transfer function of the naturally aspirated air dynamics are given in figure 3.9, where t_{na} describes the time delay of the system and τ_{na} depicts the time constant of the dynamics of air flow as well as throttle and intake valve dynamics. An identification will be performed for t_{na} and τ_{na} as a derivation from the complex physical behavior does not justify the cause. For this the ICE is exposed to steps in desired torque for various engine speeds in field tests. No significant speed dependency was determined. Therefore, the time delay will be modeled as a constant. The verification for a low and high engine speed is depicted in figure 3.10.

Turbo

State of the art engines though can produce higher torque than atmospheric pressure allows. This is performed by the process of turbocharging. The combusted gas which is pumped out of the cylinder after the ignition stroke propels a pump. This pump compresses fresh air which is then sucked into the cylinder during the intake stroke. Therefore, more torque can be produced due to a higher available air mass in the cylinder. This process is self-reinforcing: a higher intake pressure results in more combusted gas pumped out of the cylinder which yields a higher pump power compressing air to a higher degree. As it is self-reinforced the process is slow compared to the naturally aspirated path.

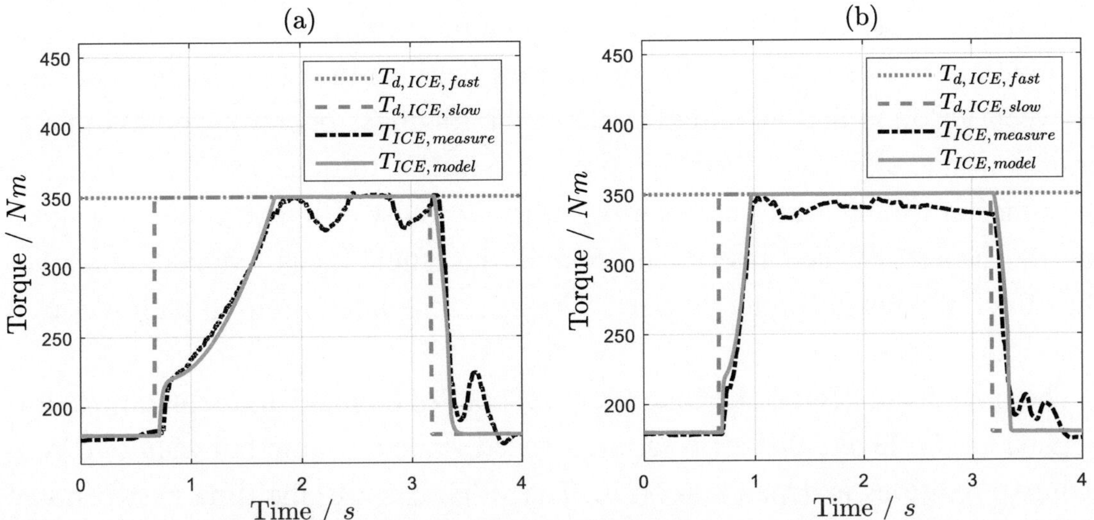

Figure 3.11.: Verification of Turbo Air Path Model - Step Responses at 1500 rpm (a) and 5000 rpm (b) of Measurement Data and Model

A physical derivation of the dynamics requires elaborte modeling of the thermodynamic and mechanical system [92]. Due to a control-oriented modeling in this work the inner processes can be neglected while still depicting the high-level input-output behavior. A convenient simplification of the turbocharged engine is a double integrator [47], as it behaves like a self-reinforcing system. A first measurement revealed a significant speed dependence which is why the proportional gains of the integrators will be depicted engine speed dependent. An identification process will be performed to fit the time constant τ_{turbo}.

Figure 3.11 shows the verification of the turbo air path model. For this, steps in desired torque have been performed at low ($1500 \frac{1}{\text{min}}$) and high ($5000 \frac{1}{\text{min}}$) engine speeds in order to verify the speed dependency. Starting at 180 Nm, still in the naturally aspirated range, a step was performed to 350 Nm at 0.7 s and back to the initial level after another 2.5 s. First a high gradient can be observed due to natural aspiration. At approximately 220 Nm the turbo dynamics take over reducing the dynamics significantly for the step at low engine speed. For high engine speed the same behavior can be observed although with a higher gradient.

Ignition Path

As already mentioned, the maximum producible torque is limited by the amount of air in the cylinder. A reduction of torque is usually performed by a reduction of air. This though, is a slow process due to the inertia of air. A faster reduction is possible by reducing the amount of gasoline injected. This method is seldomly used in state of the art engines because the stoichiometrical ratio between gasoline and air is controlled to 1 for most operating points due to emissions and efficiency. Another method to reduce torque quickly is to change the ignition timing which is usually set to the most efficient point. By igniting the air-fuel mixture later than optimal, the efficiency is reduced which has the effect of a lower torque output[3]. The gain scheduled ignition path model is depicted in figure 3.9.

The time delay depicts the time between control demand and realization. It is dependent on the crankshaft rotational speed since the time between two consecutive ignitions is dependent on it. The value is set to the time between two consecutive ignitions due to the stochastic distribution [143]. The time delay therefore is described by

$$t_{ign}(\omega_{ICE}) = \frac{n_{ign}}{n_{cyl}} \frac{2\pi}{\omega_{ICE}}, \tag{3.8}$$

---

[3] As a result the exhaust temperature rises.

where n_{cyl} represents the number of cylinders and n_{ign} the number of ignitions per cylinder per rotation where $n_{ign} = \frac{1}{2}$ for four stroke engines.

The first order low pass depicts the torque build up in the cylinder. The low-pass characteristics are described in equation 3.9. The angle in which the pressure buildup takes place is represented by $\alpha_{pressure}$. Usually full pressure builds up in approximately 60 degrees of crankshaft angle [143].

$$\tau_{ign}(\omega_{ICE}) = \frac{\alpha_{pressure}}{360°} \frac{2\pi}{\omega_{ICE}}. \tag{3.9}$$

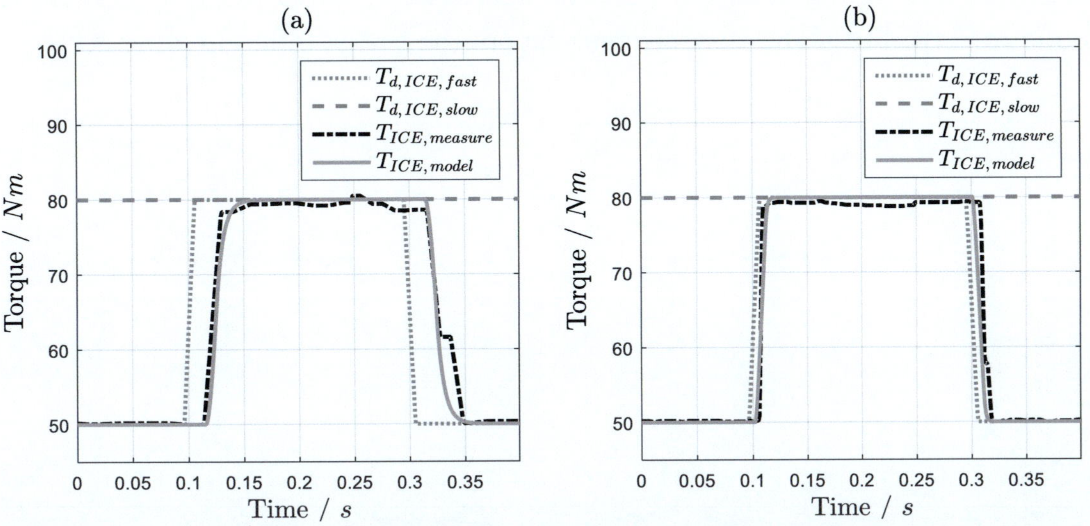

Figure 3.12.: Verification of Ignition Path Model - Step Responses at 1500 rpm (a) and 5000 rpm (b) of Measurement Data and Model

Figure 3.12 shows the verification of the ignition path model. It was performed for high and low engine speeds to verify the speed dependency. For this, the air in the cylinder was kept constant at a high value which ensures that only the ignition path is active for achieving the requested torque. This is performed by the low-level controllers regulating the ignition timing. Steps in the desired ignition path torque value have been conducted. The model dynamics show good correspondence with the measurements of the engine of the test vehicle.

3.3.4.2. Electric Motor

Compared to the complex functionality of a state-of-the-art, automotive ICE a permanent magnet synchronous electric motor (PMSM) operates simpler. The motor is comprised of a fixated stator and a pivoted rotor. The principle is based on several coils in the stator conducted by a current and therefore producing a magnetic field. Since they are supplied by a three-phase alternating current this magnetic field is rotating. The rotor consists of a permanent magnet which produces a strong magnetic field. Due to the two magnetic fields interacting with each other a torque is produced which propels the rotor. By regulating the stator current with a low-level PI controller, the torque can be adjusted. The current setpoint is given by the torque request and the operating point of the EM. For PMSMs field-oriented control is used. Again, as we focus on system-level behavior further details will not be addressed. Further information can be found in [118, 119]. The system-level data of the EM is found in table 3.1.

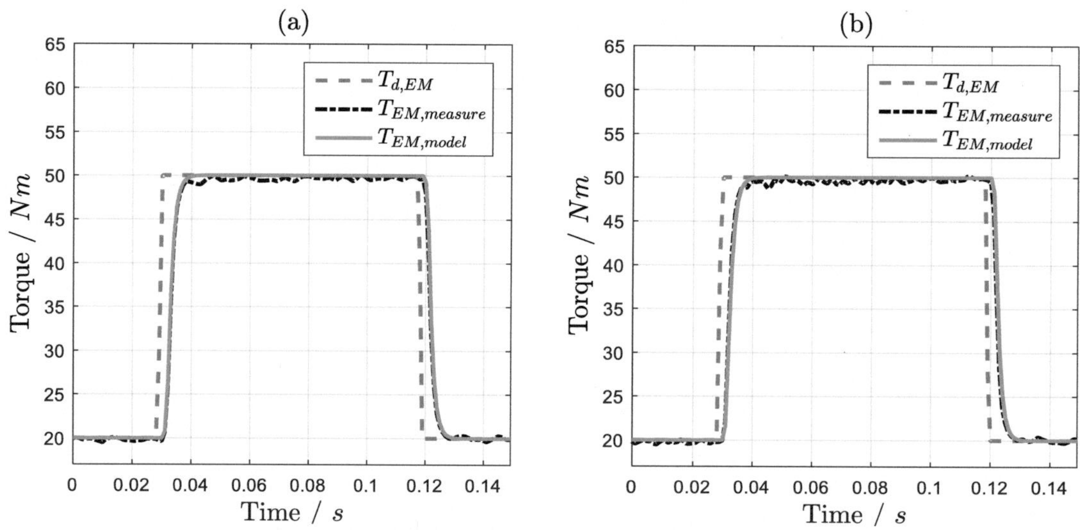

Figure 3.13.: Verification of Electric Motor - Step Responses at 1500 rpm (a) and 5000 rpm (b) of Measurement Data and Model

This assembly results in a highly linear behavior with high dynamics and low latencies compared to an ICE. This can be traced back to the system only consisting of electronic and electric components which are driven by fast current dynamics.

The EM will therefore be modeled as a linear first order low pass with a time delay including a rate limitation of the torque. The low pass filter hereby depicts the current control dynamics whereas the time delay models the communication delays within the electronic component. The only major nonlinearity in the system is given by a limitation of the current change which is represented by the rate limiter in the model. The parameters are constant and independent of other states such as rotor speed or applied torque. The only limiting factor remains the maximum torque which is dependent on the rotor speed due to field weakening.

An identification has been performed to determine the values of the time delay and the time constant of the first order low pass. For this, the EM was exposed to steps in desired torque while the resulting actual torque was measured. Steps of 30 Nm starting at 20 Nm as a positive step and at 50 Nm as a negative step were performed. This was done at two different rotational speeds in order to identify speed dependencies. A low speed of $1500\,\frac{1}{\text{min}}$ and a high speed of $5000\,\frac{1}{\text{min}}$ were chosen. The results are depicted in figure 3.13. No speed dependencies were identified. The model represents the actual behavior of the system to a high degree of accuracy for positive as well as negative steps. The parameters of the rate limitation are derived from design specifications of the EM.

3.3.5. Control Network

Three ECUs and two bus systems are involved in the traction control communication structure of the investigated vehicle. As ECUs operate discretely with a distinct sample time for the upper-level controls they will also be modeled as time discrete systems. The DECU and EECU use a sample time of 1 ms for the upper-level controls. The CECU has a sample time of 10 ms for the upper-level torque control. This is due to the lower dynamics of the actuator. The main controls are located on the CECU. Here, the information from the accelerator pedal position is evaluated. This generates the driver torque request which is then allocated to the individual actuators. The detailed algorithm used in the vehicle will be described in chapter 5. The main controls of this system have been modeled to meet the same behavior as the actual function implemented in the ECU.

Due to several functional layers and function calls within an ECU, communication delays occur. Also, the arbitration of messages with non-synchronized receivers produces a delay. This occurs by communicating signals via a bus to another ECU. For this reason, the communication between ECUs will be time delayed. Usually this delay is stochastically distributed over a small spectrum. In order to achieve reproducible and therefore, comparable simulation results, the bus delay times will be chosen constant with their distinct delay times to be identified.

In order to determine the unknown delay times of the bus communication an identification is performed. A cross correlation as described in [142] was used.

For this, a signal which was generated in one ECU and sent to another ECU has been measured in both ECUs with a single measurement setup. This ensures time synchronicity. This was performed for both bus systems as they use different technologies and have different workloads which limits the bandwidth of the system. For each, ten sampling intervals during various maneuvers have been chosen to determine the time delay.

Additionally, the bandwidth of a CAN bus communication is limited. Therefore transmitted signals are quantized. Depending on the required resolution signals are quantized between 12 and 16 bit. The same quantization used in the vehicle communication will be implemented in the simulation model.

3.4. Open-Loop Verification of the Vehicle Model

In the previous section all individual model components were described and unknown parameters identified. In this section an open-loop verification of the complete vehicle model is conducted. In order to decouple the plant from the controller which has not yet been addressed the system is excited with synthetic desired torques. This also allows the verification of the tire model as it can be run for a wide range of slip values incorporating the pseudo-linear as well as the nonlinear part without controller interference.

The experiment is conducted on a snow covered even road in second gear with the described test vehicle. Starting in a stable dynamic vehicle state in longitudinal direction low torque steps from 50 to 80 Nm and back are applied to the ICE. This leads to a loss of traction as the wheels start to spin. High oscillatory behavior of the driven wheels and especially the longitudinal wheel slip can be observed. Reducing the torque leads to a decrease in rotational speeds and reentering a stable vehicle condition. Due to the torque excitation and leaving the stable driving condition drivetrain oscillations occur. These are more pronounced at lower speeds.

The model is excited with the same steps in desired engine torque. The results of the model verification are depicted in figure 3.14. The figure shows measured signals of the experiment comparing them to the corresponding signals of the simulation. The model is an accurate representation of the actual system. This can be drawn from the reaction of torque input to speed and wheel slip output. The oscillatory behavior is depicted to a high degree of accuracy regarding frequency and also amplitude. Also the decreasing oscillation intensity with higher

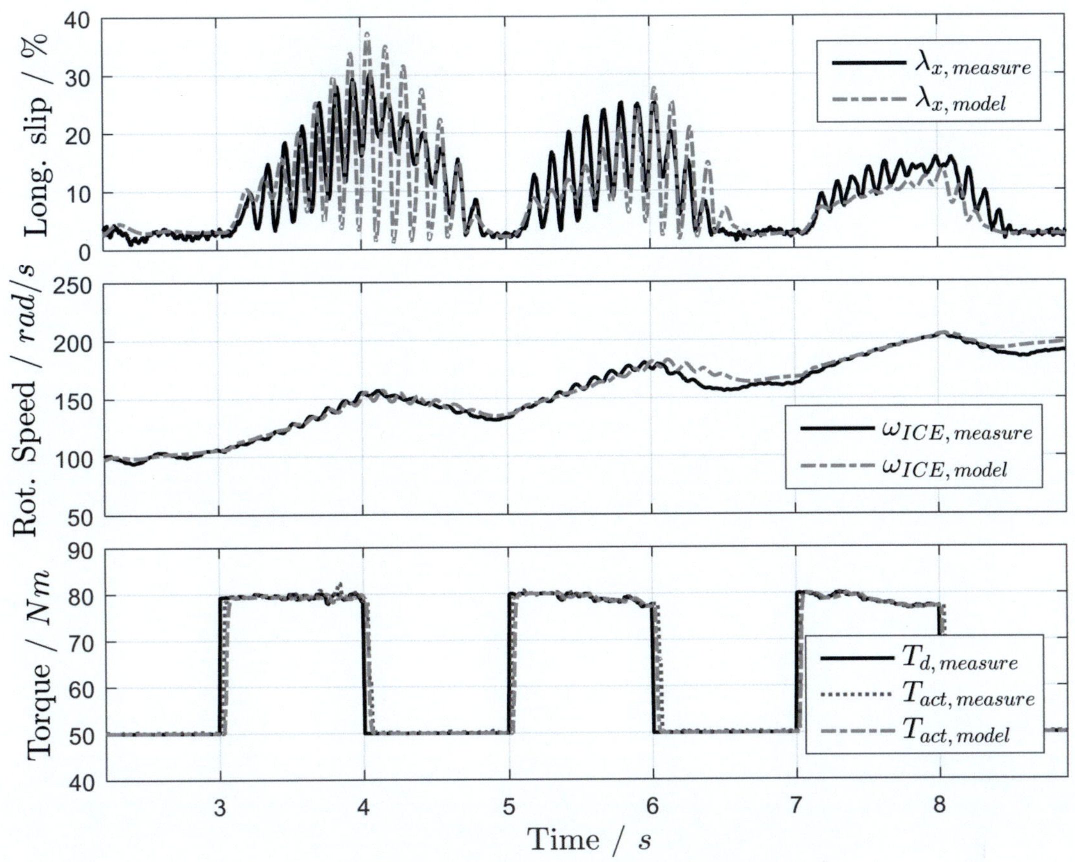

Figure 3.14.: Verification of Vehicle Model

speeds is depicted accurately. The actuator model follows the measured signal to a high degree of accuracy.

Stochastic variations in road unevenness or variations in the friction coefficient were not modeled. This leads to different amplitudes in drivetrain oscillations as these excitations are not present in the model. Another limitation of this comparison is the difference in final rotational speeds of the engine speed although starting from the same values. The reason for this is an open-loop integral. Unknown disturbances act on the system and also the ICE torque is not depicted perfectly which leads to an offset over time.

4 Control Design

This chapter deals with the synthesis of speed-based traction controllers. First, general requirements on TC systems and quantitative design limitations will be stated. Then, the controller setup using the nonlinear synthesis method IOL to the present control problem will be shown. Thereafter the oscillatory plant behavior will be discussed which motivates the need to distinguish between slow and fast actuators when designing a traction controller. Based on this study two controller designs will be presented. One is applied to systems with low actuator dynamics, the other, to systems with high actuator dynamics incorporating further control objectives. The chapter closes with a closed-loop simulation of both designs.

4.1. Requirements for Traction Control Systems

First, requirements for the final design must be addressed in order to develop a suitable control algorithm. In general, requirements involve the functionality itself, address different users, and are defined by various instances. These involve the car manufacturer, independent standards, and governmental regulations. Users can be the final customer but also the engineer developing the function. Requirements on the functionality define the properties of the system to be developed.

Since the function to be developed supports the driver an application and user-oriented viewpoint will be considered. The requirements are categorized into three major groups as seen below. They were gathered from literature [44] as well as workshops with experts on driving dynamics development.

The first group describes the functionality of traction controllers. It includes which overall tasks must be fulfilled by the controller. The most important feature is that the controller prevents vehicle instability and a loss of maneuverability. These functions must be available for changing vehicle parameters such as mass through passengers and payload. Additionally, the controller must provide

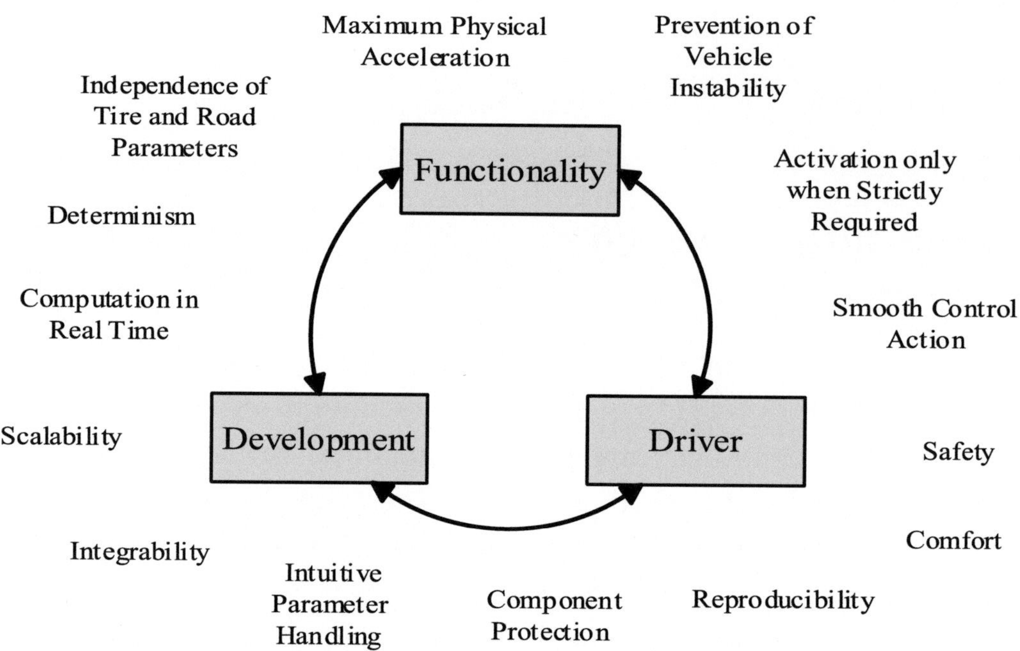

Figure 4.1.: Requirements of TC in Automotive Applications

high levels of performance for varying tire properties such as dimensions and resiliency. Also, an independence of road conditions must be guaranteed. When they are unknown the controller still needs to function properly and ensure vehicle stability. Stability may be achieved easily by reducing torque tremendously obtaining only low slip values. This though interferes with the last requirement of achieving the maximum physical acceleration. Since the driver misjudges the road conditions and desires a higher acceleration that can physically be achieved his demand must be met as well as possible. These requirement characteristics are solely defined by the traction controller itself.

After discussing general requirements, strict requirements for the controlled system will be defined. These were discussed thoroughly in workshops with experts on this subject. An important variable is the overshoot time. It defines how long the vehicle is in an unstable state due to excessive wheel slip. Another variable concerns damping of the controlled system. The objective is to achieve a quick rising time but also a low oscillatory behavior. Therefore, the maximum settling time was defined to be $t_{settle,max} = 0.4$ s and the minimum damping of the system to be $\zeta_{min} = 0.6$. Another quantitative requirement is the minimum torque output of the traction controller. Blocking of the wheels must be prevented in any case which is why the torque output is limited to 0 Nm.

The second group of requirements regards the development process of the traction controller. Vehicle manufacturers produce many different vehicles with various powertrain configurations. For this reason, the developed system must be applicable for all configurations addressed by the manufacturer. This reduces complexity and keeps further developments to a minimum. The next requirement addresses the implementation of a controller to the existing architecture. In practical terms this means that the controller can only use information based on existing sensors and models to reduce complexity and costs. This ensures seamless integration. Further gains in complexity reduction can be achieved by an increased focus on parameter handling. With the use of physical models, physical parameters are employed. These can be easily obtained through dimensioning information. It also addresses the ease of adjusting the parameters of the system for the control engineer to find optimal values in all driving maneuvers. This must be taken into account while designing the control system. The parameterization of the control designs in this work will serve as initial values for tests and for an analytic description only. It still needs to be adjusted by the control engineer in driving tests. Another important requirement is component protection. A damaging of the drivetrain or vehicle by the traction controller must be prevented at all times. A control system must also be applicable to ECUs. Only controller structures with a computation complexity that can be handled by industrial ECUs are to be considered. Closely related is the requirement of determinism which must be guaranteed for diagnostic purposes.

The last group of requirements addresses the customer or the driver of the vehicle. The most important is safety which must be ensured for driver and passengers at all times and for all operating conditions. One example is that during operation the traction control system might produce more torque than the driver requested which may lead to serious accidents. This and other failures must be avoided. Furthermore, during operation of the traction control system discomfort must be suppressed. This may lead to exhaustion and disturbs the driver as well as passengers. The last requirement addresses the behavior of the function. It must be reproducible, intuitive and comprehensible for the driver. This ensures acceptance by the driver and therefore an understanding of the functionality. These requirement characteristics are defined by the traction controller as well as the integration to the existing architecture.

State Machine

To meet the requirement of the controller only being active when strictly needed the activation and deactivation behavior must be defined. If the combined torque exerted on the wheels demanded by the driver T_{driver} exceeds the traction limits of tire road conditions TC must activate. The condition for this is given by

$$\lambda_x \geq \lambda_r, \tag{4.1}$$

where λ_x is the actual longitudinal slip and λ_r the reference slip. The traction controller then must limit the desired driver torque to ensure minimal deviation between actual and reference slip. This must only be done if strictly required. Pinto [103] therefore defines the deactivation condition as a hysteresis

$$\lambda_x < K\lambda_r, \tag{4.2}$$

with a scaling factor $K \leq 1$ to prevent activation toggling. Experiments though have shown that this condition is not sufficient to ensure the requirement of smooth control action. When large slip undershoots occur the controller deactivates. The torque request then switches abruptly back to the driver's request. This can cause the slip again to exceed the desired value which in turn activates the controller. These conditions therefore, can create a limit cycle. For this reason, another condition will be added which must apply simultaneously for deactivation:

$$T_{controller} \geq T_{driver}. \tag{4.3}$$

The controller torque $T_{controller}$ must exceed the driver torque demand. This prevents steps in desired torque ensuring a smooth transition from TC back to the driver torque demand.

4.2. Controller Structure

First, a brief discussion on a suitable design method will be carried out. Known publications on speed-based TC propose linear PI controllers for the inner control loop. Nonlinear controllers have not been applied. Performant nonlinear designs though have been thoroughly discussed on direct slip TC. Noticeable designs are SMC, feedback linearization and flat control. Although offering precise control performance, flat control is considered overly complex in experimental tuning. SMC is a robust nonlinear control method but is prone to produce a chattering control signal. A promising control design is feedback linearization or input-output linearization. It offers high control performance with the ease of integration and experimental tuning [23]. Therefore, input-output linearization will be used in the present work to design a controller. The following sections are based on the publication [142].

The structure of speed-based TC with a controller design based on IOL is depicted in figure 4.2. The IOL and plant together form a linear system from its input w to output y. The output is fed back and compared to the reference value r. Its control error is input to the linear PID controller. The nonlinear controller itself consists of the linear PID controller and the IOL term. Notice that the input of the plant is provided by the IOL which is driven by the controller output and state variables of the plant. The reference wheel speed ω_r is generated by the outer-loop controller consisting of an estimation for the reference slip value λ_r and a conversion. It generates the reference signal from the reference slip and vehicle speed v_x.

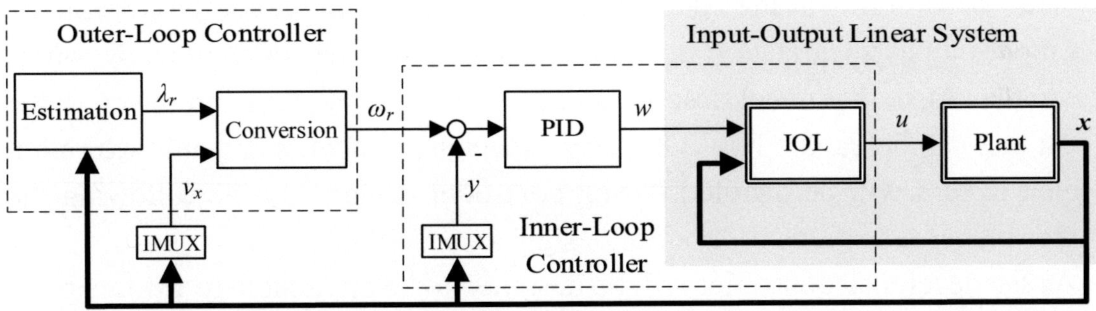

Figure 4.2.: Controller Structure of Speed-Based Traction Control using Input-Output Linearization

As already discussed in the introduction the setpoint, desired slip value or reference value defines the overall system behavior. If chosen too low acceleration potential is lost and if chosen too high tractive forces drop significantly reducing traction potential and leaving the vehicle unstable during cornering. Road parameters are mostly unknown and can vary quickly. Therefore, an extensive estimation must be performed. This though, is not part of this work. References on this matter can be found in the state of the art 1.2.1.

For this work it is assumed that a desired slip value as well as the vehicle velocity is provided to a reasonable degree of accuracy. In order to generate a setpoint for the described controller structure the slip definition in equation 1.1

is considered. Reformulating this equation given the longitudinal vehicle speed v_x, the dynamic wheel radius r_{whl}, and the reference slip value λ_r yields the following expression of the reference wheel speed

$$\omega_r = \frac{v_x}{r_{whl}(1 - \lambda_r)}.\tag{4.4}$$

In the following sections a nonlinear controller will be designed for plants with slow and fast actuators. First the IOL must be performed followed by the design of a linear PID controller. Note that the designs are valid for fixed gears.

4.3. On the Choice of Design Models

The choice of the design model is not only to be based on the dominant dynamics of the system but also on the dynamics of the actuator involved and the design method. Simple output feedback systems may be designed for high order systems, methods that utilize system states in the feedback loop require systems of low order as the system states need to be available for the controller[1]. As input-output linearization makes use of state variables a low order system is desirable. Also, the model described in the previous chapter is not given in the form that can be handled by IOL. This especially concerns actuator saturation which in turn will be handled during controller design. In the following the minor model reductions for control design are described.

As the developed controller will be designed to be applicable to various powertrain configurations, a SISO system is considered. Therefore, the present actuator redundancy in parallel HEVs needs to be resolved. This is done by forming the transfer function of the combined actuator system. As only a rough depiction for the actuator dynamics is preferred due to the applicability of IOL the combined transfer function will be reduced to a first order low pass. For details refer to sections 5.1 and 5.2.

A further reduction concerns dynamic axle load distribution. It will not be incorporated in the design model as it does not affect the governing nonlinearity of the system. Pinto et al. compare various design models for TC discussed in literature. None of the depicted models make use of dynamic axle load distribution. It is on the other hand valued for simulation purposes increasing model accuracy [103].

---

[1]Either by measuring or observing

Drivetrain Dynamics

As pointed out by many authors the oscillatory behavior of the drivetrain negatively influences traction control performance [16]. The verification of the test vehicle model in figure 3.14 shows high oscillatory behavior when exceeding the critical slip which therefore must be addressed. Active damping is an effective and widely used control scheme to deal with this issue [111, 136]. Known works though either do not incorporate actuator dynamics to damping control design or assume an actuator with high dynamics such as an EM. When applying active damping to a plant with slow actuators with time delays the system damping does not increase. In contrast the system damping only decreases until becoming unstable as the following example with test vehicle parameters illustrates.

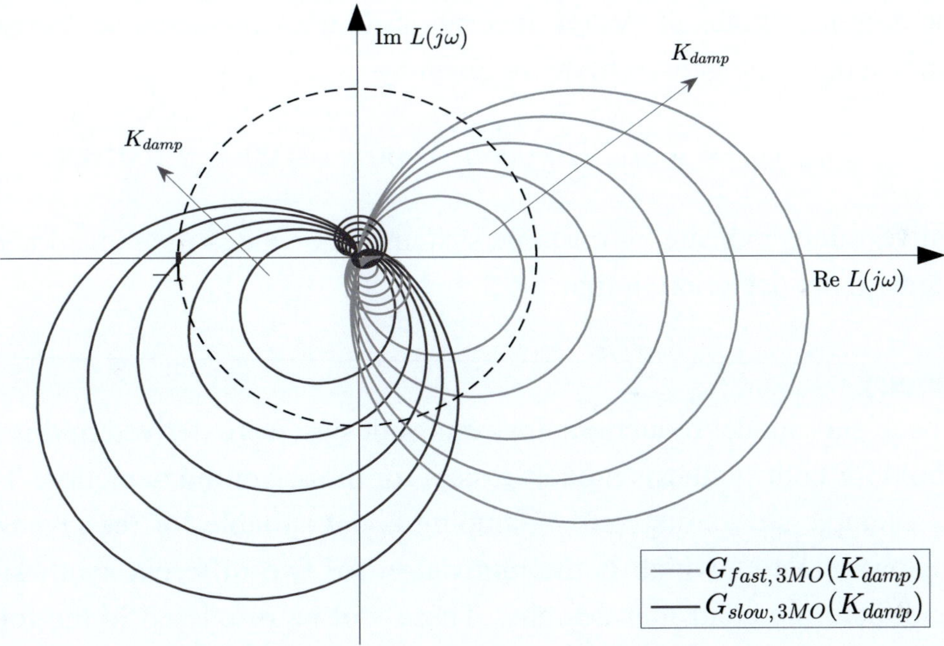

Figure 4.3.: Nyquist Plot for Active Damping Control of a Three-Mass Oscillator for Two Different Actuator Dynamics at Different Values of Controller Gain

In figure 4.3 the Nyquist plot of two open-loop damping controller designs are shown. Both controlled systems consist of the three-mass oscillator introduced in section 3.3.3. The same linearization for slip and tire force at maximum traction is applied. The first system $G_{slow,3MO}$ makes use of the resulting actuator model of Design A whereas system two $G_{fast,3MO}$ uses the resulting actuator model of Design B^2. The integrated damping controller regulates the difference in rotational speeds of engine and wheels[3] as suggested by [111, 136, 103]. The factor of the damping controller is given by $K_{damp} = [0.7, 1, 1.3, 1.6, 1.9]$. The plot shows that the system using the fast actuator remains stable with high stability margins with increasing controller gain. The system with the slow actuator becomes unstable using the same controller gains.

The damping coefficients of the dominant poles of the system using the fast actuators on the other hand rise with increasing gain

$$\zeta_{fast,3MO} = [0.0520, 0.0626, 0.0732, 0.0838, 0.0944]. \qquad (4.5)$$

Increasing the gain further results in higher damping. The maximum damping coefficient is $\zeta_{EM,3MO,max} = 0.352$ at $K_{damp} = 10.4$. The damping coefficients of the dominant poles of system incorporating the slow actuator decreases for the shown damping gains which are given by

$$\zeta_{slow,3MO} = [0.0127, 0.0066, 0.0007, -0.0052, -0.0109]. \qquad (4.6)$$

Negative values indicate an unstable system. The values mentioned above coincide with the depiction in figure 4.3.

Summary

The necessary model reductions for control design were derived in this section and hold for both synthesis models except for the drivetrain depiction. The previous example shows that active damping is not suitable for the given system with slow actuators which is the motivation for two different synthesis models and therefore controller designs. These will be discussed in the following starting with a novel traction control method for an oscillatory drivetrain being propelled by actuators with a low dynamic response and time delay.

---

[2] Refer to sections 5.2.3 and 5.1.3 respectively for the derivation of the actuator models
[3] Which are scaled to crankshaft level

4.4. Control Design for Slow Actuators

The chosen design method will now be applied to designing a speed controller for TC. The first controller is designed for plants with slow and time delayed actuators. First, the choice of the design model is argued. Thereafter a controller will be designed using the method of input-output linearization combined with a feedback of the output variable regulating the desired drivetrain speed with a PID controller. Finally, an observer will be presented estimating a virtual rigid drivetrain speed which is required for the chosen design model. Changes in the drivetrain configuration during hybrid and electric drive will be considered.

4.4.1. Design Model

As shown in the previous section an active damping approach suggested by the literature to enhance control performance cannot be applied to systems with slow actuators. Therefore, a unique approach is chosen to prevent conservative parameterization and therefore performance reduction of the controlled system. A rigid drivetrain with no oscillatory behavior is assumed as a design model for the controller. As the system does not consist of a rigid drivetrain, this virtual state must be observed which will be described in 4.4.3.

After arguing the motivation for system order reduction, the assumed nonlinear design model for TC concerning slow actuators is given by

$$
\dot{\boldsymbol{x}} = \begin{bmatrix} \dot{x}_1 \\ \dot{x}_2 \\ \dot{x}_3 \end{bmatrix} = \begin{bmatrix} \dfrac{1}{\tau_{red,slow}}(u - x_1) \\ \dfrac{1}{J_{red}}\left(x_1 - \dfrac{r_{whl}}{i_{tot}}(F_x - F_r)\right) \\ \dfrac{F_x - F_r - F_a}{m_{veh}} \end{bmatrix},
$$

$$
y = \frac{x_2}{i_{tot}},
$$

(4.7)

where x_1 is the actual engine torque. The drivetrain speed at crankshaft level and the vehicle velocity are represented by x_2 and x_3 respectively.. The time constant of the reduced actuator system depicted as a first order low pass is described by $\tau_{red,slow}$. It takes on different values, $\tau_{red,slow,hyb}$ for hybrid and $\tau_{red,slow,ed}$ for electric drive, compare section 5.1.

The inertia of the drivetrain reduced to crankshaft level is described by J_{red}. The method of Laschet introduced in section 3.3.3 does not provide instructions for reducing a torsional oscillator to a rigid system. Therefore, the sum of all inertias will form the reduced total drivetrain inertia. Using the drivetrain configuration in figure 3.6 the parallel axis theorem by Steiner is applied [27]. The reduced inertias are given by

$$J_{red,hyb} = J_{ICE} + J_{EM} + \frac{J_{diff}}{i_{gb}^2} + \frac{J_{whl}}{i_{tot}^2} + \frac{J_{belt}}{i_{tot}^2}, \tag{4.8}$$

for hybrid drive, where for electric drive

$$J_{red,ed} = J_{EM} + \frac{J_{diff}}{i_{gb}^2} + \frac{J_{whl}}{i_{tot}^2} + \frac{J_{belt}}{i_{tot}^2}. \tag{4.9}$$

holds due to the decoupling of the crankshaft.

4.4.2. Linearization of Plant Dynamics

The design model presented in equation (4.7) is an input affine nonlinear system which can be linearized by IOL. To apply this synthesis method first, the system output needs to be considered. It is given by:

$$y = \frac{x_2}{i_{tot}}. \tag{4.10}$$

Now the Lie derivatives of the system output equation are determined until a dependency on the input is given. The first derivative of y of the analyzed system is

$$\dot{y} = \frac{\dot{x}_2}{i_{tot}}. \tag{4.11}$$

Inserting $\dot{x}_2$ from equation (4.7) yields

$$\dot{y} = \frac{\frac{1}{J_{red}}\left(x_1 - \frac{r_{whl}}{i_{tot}}(F_x - F_r)\right)}{i_{tot}}$$

$$= \frac{1}{J_{red} i_{tot}}\left(x_1 - \frac{r_{whl}}{i_{tot}}(F_x - F_r)\right). \tag{4.12}$$

Now the tire force F_x occurs in the linearizing equation. Since it cannot be measured and only be estimated with much effort, the dependency needs to be eliminated from the linearizing law by replacing it with measurable variables. Here the third state equation in (4.7) is useful which forms to

$$
\begin{aligned}
F_x &= m_{veh}\dot{x}_3 + F_r + F_a \\
&= m_{veh}\dot{x}_3 + F_r + \frac{1}{2}\rho c_x A x_3 |x_3|
\end{aligned}
\tag{4.13}
$$

The mass of the vehicle can be estimated online to the required degree of accuracy and the acceleration $\dot{x}_3$ and vehicle speed x_3 are already measured. After canceling out the rolling resistance F_r the derivative of y is given by

$$
\dot{y} = \frac{1}{J_{red}i_{tot}} \left(x_1 - \frac{r_{whl}}{i_{tot}} \left(m_{veh}\dot{x}_3 + \frac{1}{2}\rho c_x A x_3 |x_3| \right) \right) .
\tag{4.14}
$$

Since the input u is not part of the equation the second derivative is determined

$$
\begin{aligned}
\ddot{y} &= \frac{1}{J_{red}i_{tot}} \left(\dot{x}_1 - \frac{r_{whl}}{i_{tot}} \left(m_{veh}\ddot{x}_3 + \frac{1}{2}\rho c_x A \left(\dot{x}_3 |x_3| + \frac{x_3^2}{|x_3|} \right) \right) \right) , \\
&= \frac{1}{J_{red}i_{tot}} \left(\frac{1}{\tau_{red,slow}}(u - x_1) - \frac{r_{whl}}{i_{tot}} \left(m_{veh}\ddot{x}_3 + \frac{1}{2}\rho c_x A \left(\dot{x}_3 |x_3| + \frac{x_3^2}{|x_3|} \right) \right) \right) .
\end{aligned}
\tag{4.15}
$$

The second derivative of the system output yields a dependency upon the input. Since the order is lower than the rank of the system dynamics non controllable internal dynamics exist. A proof of the stability has been proposed by Reichensdörfer et al. for a 5-DOF system [105].

In the final equation (4.15) there is a dependency on the derivative of the vehicle acceleration. Since this signal is not measured it will be calculated by determining the derivative of the acceleration signal. Due to the prevalence of noise the time constant of the differentiator will be set to $\tau_D = 0.1s$. Applying equation (2.19) to our system yields the differential equation of second order of the reference model

$$
\ddot{y} + a_1\dot{y} + a_0 y = Vw .
\tag{4.16}
$$

Inserting equation (4.15) into previous equation and then solving for u yields the new control input w for the plant (4.17). This leaves three variables to be specified, a_1, a_0 and V which define the dynamics of the resulting linear system as the reference model.

$$u = x_1 + \tau_{red,slow}\left(J_{red}i_{tot}(Vw - a_1\dot{y} - a_0 y)\right.$$

$$\left. + \frac{r_{whl}}{i_{tot}}\left(m_{veh}\ddot{x}_3 + \frac{1}{2}\rho c_x A\left(\dot{x}_3|x_3| + \frac{x_3^2}{|x_3|}\right)\right)\right). \tag{4.17}$$

4.4.2.1. Reference Model

As pointed out in section 2.3.4 performing an input-output linearization leaves the linear input-output behavior to be specified by choosing the polynomial co-efficients. It is limited though to the order of the external dynamic. In the present case the order is two. The motivation for the choice of the parameters is to approximately meet the physical behavior of the system. This makes adjusting the parameters in experiments easier and more intuitive. This is an important feature as the controller is applied to a wide range of vehicles causing many parameter variations to ensure satisfactory performance in each case. Focusing on the drivetrain, the system behavior is roughly represented by a first order lowpass and an integral element. The first order lowpass depicts the engine dynamics and the integrator the behavior from torque to rotational speed. All parameters are known and have been introduced in chapter 3.3 and earlier in this section. A similar approximation for the reference model was introduced by Reichensdörfer et al. [105]. Here, only the wheel inertia was taken into account, whereas this work takes the entire drivetrain inertia into account depicting the synthesis model more accurately.

The transfer function is given by

$$G_{ref} = \frac{1}{\tau_{red,slow}s + 1}\frac{1}{J_{red}}\frac{1}{s}\frac{1}{i_{tot}} = \frac{\dfrac{1}{\tau_{red,slow}J_{red}i_{tot}}}{s^2 + \dfrac{1}{\tau_{red,slow}}s}. \tag{4.18}$$

Since the gear ratio is included in equation (4.18) there is a dependency on the selected gear. This would require the design of several controllers if included as a dynamic state. For the reference model though a fixed gear ratio is chosen to achieve only one single model independent from the selected gear. Therefore, one single controller can be designed which is applicable to all gears since the linear plant behavior does not vary with the gear ratio. For this problem, the second gear is chosen. This is motivated by the fact that the first is not considered a driving gear and only used for the start. The second gear usually still provides enough torque on the axle for the wheels to skid but has a wide operating range in terms of vehicle velocity.

By equating (4.16) with (4.18) the parameters of the reference model are determined directly and are given by

Table 4.1.: Parameters of Reference Model

Parameter	Equation
V	$\dfrac{1}{\tau_{red} J_{red} i_{tot}}$
a_1	$\dfrac{1}{\tau_{red}}$
a_0	0

4.4.3. Observer Design

The control design model assumes a rigid drivetrain, the real system though shows pronounced oscillatory behavior. A dominant pole has been eliminated during model order reduction. This circumstance would lead to a negative effect on control performance due to conservative calibration. The reason for this lies in the calculation of high control deviations when using the measured motor speed due to its oscillatory behavior. To overcome this issue the rigid drivetrain speed must be made available to the controller. It coincides with the measured motor speed for low frequencies but does not show oscillations for higher frequencies in the range of the eigenfrequencies of the drivetrain.

Low pass filtering of the oscillatory motor speed does not fulfil the control requirements as it produces a high phase delay also resulting in conservative calibration. This is due to filtering all frequencies equally below its crossover frequency. A different design, only filtering desired frequency ranges is a band stop or notch filter. These can be applied when there is a distinct frequency that needs to be filtered. In the present nonlinear problem this is not a trivial task.

As described in section 3.3.3, the drivetrain eigenfrequencies shift depending on the actual slip. Therefore, a notch filter cannot be applied with reasonable effort.

Since a feed forward filter does not achieve the goal of providing a rigid drivetrain speed free of phase delay, an observer will be introduced to observe the virtual rigid drivetrain speed. Observers were developed to observe unmeasurable states of a controlled system by using measurable states and estimating the unknown by an internal feedback loop. These can then be used for control design of state feedback controllers.

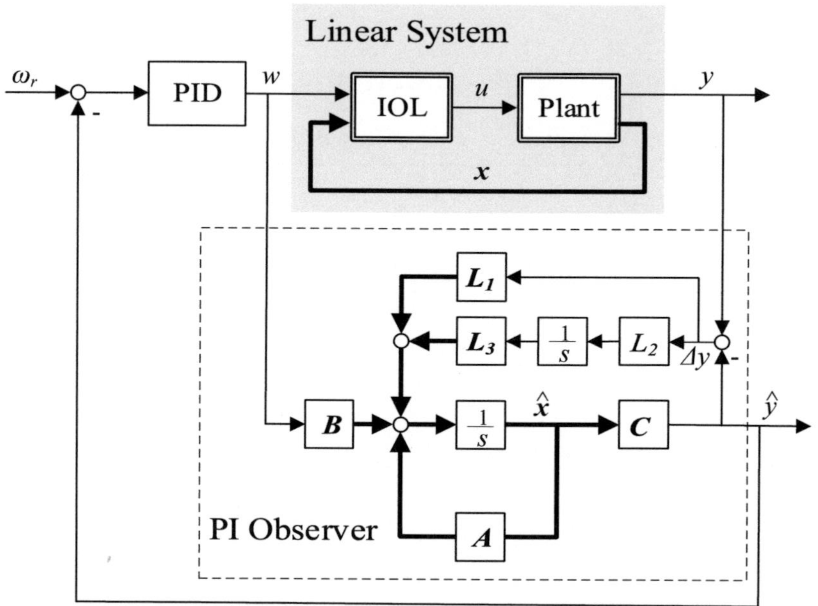

Figure 4.4.: Controller Structure with PI Observer

Various observers have been proposed in literature. A well-known and robust approach is the Luenberger observer [85, 145]. More sophisticated approaches are Kalman filters. These though have a higher computational as well as parameterization effort and are often used when high stochasitc influences such as noise are present [127, 145]. For this reason, Luenberger-type observers will be selected for the present work. These though leave a steady state error as the feedback loop only incorporates a proportional gain. This is critical in slip control as small deviations from the reference value can dramatically change the plant behavior. A variation of the Luenberger observer has been introduced which adds an integral part to the observer feedback [14, 19, 124]. With this integral feedback the error dynamics converge to zero. This concept will be used in this work to observe a virtual rigid drivetrain speed. The implementation to the control structure is depicted in figure 4.4. Two systems are excited by the same input. Notice that the observer is excited with the output w of the PID controller. The plant on the other hand receives the input u from the IOL. The output of the real system y is compared to the output of the observer system $\hat{y}$. The difference Δy is multiplied with the proportional observer vector $\boldsymbol{L_1}$ and fed back[4]. Additionally, the integral of $L_2\Delta y$ is multiplied by $\boldsymbol{L_3}$ acting as a further observer system input. The observed output is used as the controlled variable and compared with the reference signal ω_r.

The dynamic equation of the proportional integral (PI) observer is given by

$$\dot{\hat{x}} = A\hat{x} + L_3\hat{f}Bu + L_1\Delta y$$
$$\dot{\hat{f}} = L_2\Delta y,$$
(4.19)

with

$$\Delta y = y - \hat{y} = Cx - C\hat{x}.$$
(4.20)

This gives the state space expression of the observer

$$\begin{bmatrix} \dot{\hat{x}} \\ \dot{\hat{f}} \end{bmatrix} = \underbrace{\begin{bmatrix} A - L_1C & L_3 \\ -L_2C & 0 \end{bmatrix}}_{A_e} \begin{bmatrix} \hat{x} \\ \hat{f} \end{bmatrix} + \begin{bmatrix} B \\ 0 \end{bmatrix} u + \begin{bmatrix} L_1 \\ L_2 \end{bmatrix} y.$$
(4.21)

The system matrices A, B and C are defined as the reference model introduced in 4.4.2. The estimation error of the observer is defined as $e(t) = \hat{x}(t) - x(t)$, and yields error dynamics of

$$\begin{bmatrix} \dot{e} \\ \dot{\hat{f}} \end{bmatrix} = A_e \begin{bmatrix} e \\ \hat{f} \end{bmatrix},$$
(4.22)

The observer design task is to define $\boldsymbol{L_1}$, $\boldsymbol{L_2}$ and $\boldsymbol{L_3}$ so that desired error dynamics are yielded. By determining the characteristic polynomial (4.23) of the system the poles can be placed in desired locations. Here, λ specifies the pole or eigen value and $\boldsymbol{I}$ describes an eye matrix.

---

[4]This is the only feedback loop of the Luenberger observer

$$\det(\lambda \boldsymbol{I} - \boldsymbol{A}_e) = 0, \qquad (4.23)$$

For this third order system a two-step design process is used. First, the desired error dynamics of the Luenberger observer are achieved by tuning the proportional gain, then the integral gain will be added to achieve further requirements, similar to the Ziegler-Nichols parameterization of PI controllers [87].

A Luenberger observer is usually designed to be 2-6 times faster than the observed system with lower values when much noise is present [86]. As the objective of the observer in this work differs from usual applications, the following design requirements are specified:

1. The observer dynamics must be smaller than the drivetrain dynamics to achieve a filter effect: $\omega_{0,obs} < \omega_{0,dt}$.

2. The observer dynamics must be similar to the dynamics of the reference model to achieve low phase delays: $\omega_{0,obs} \approx \omega_{0,ref}$.

3. High damping must be achieved as oscillatory error dynamics are not desirable.

4. The error must fall below 1% after maximum settling time for a step response to meet the requirements defined in 4.1.

Step 1

To define the poles of the Luenberger observer, first the characteristic polynomial must be calculated by

$$\det(\lambda \boldsymbol{I} - (\boldsymbol{A} - \boldsymbol{L_1} \boldsymbol{C})) = 0. \qquad (4.24)$$

As the matrix $\boldsymbol{A}$ is 2×2 this yields a characteristic polynomial of order two

$$\lambda^2 + 2\zeta \omega_{0,obs}\lambda + \omega_{0,obs}^2 = 0. \qquad (4.25)$$

Now damping ζ and eigenfrequency ω_0 of the system can be chosen freely considering the requirements. Since oscillatory behavior is undesired the damping coefficients are defined as

$$\zeta_{hyb} = \zeta_{ed} = 1. \qquad (4.26)$$

The eigenfrequency of the observers are defined as

$$\omega_{0,obs,hyb} = 0.7\,\omega_{0,ref,hyb}\,,$$
$$\omega_{0,obs,ed} = 1.2\,\omega_{0,ref,ed}\,.$$

$$(4.27)$$

The factors of the observer eigenfrequency are chosen differently due to a different oscillatory behavior during hybrid and electric drive. During hybrid drive the two-mass flywheel adds an additional DOF to the system which is handled with a lower factor. Inserting equation (4.25) into equation (4.24) determines the parameters of L_1. With step 1 requirements 1-3 are addressed.

Step 2

After defining the behavior of the proportional feedback, the Luenberger observer, now the additional integral gains L_2 and L_3 need to be specified. Due to a parametric redundancy $L_2 = 1$ is chosen arbitrarily. Then, the matrix L_3 is defined to meet the fourth requirement regarding the error dynamic. The results are depicted in figure 4.5. It shows the responses of the speed deviattion of measured y and estimated motor speed $\hat{y}$ for hybrid drive (a) and electric drive (b) when exerting a step on the measured motor speed. For each mode

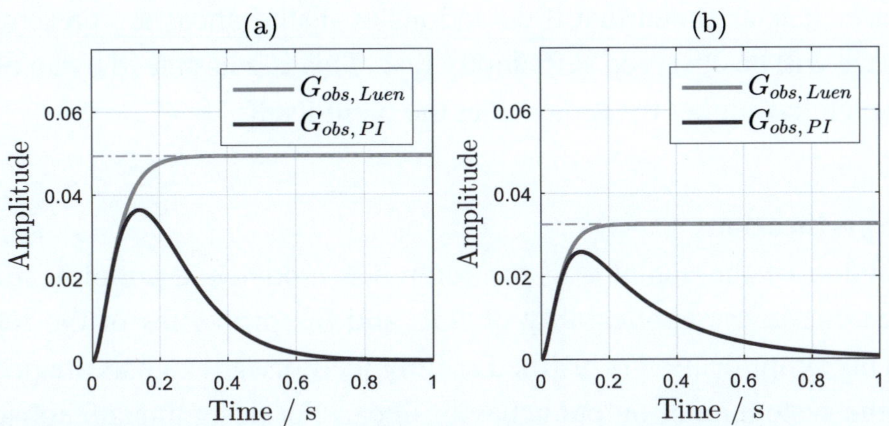

Figure 4.5.: Step Responses of Observer Error Dynamic, Hybrid (a) and Electric Drive (b)

the responses of the Luenberger Observer and the PI Observer are shown. The Luenberger observer leaves a remaining error amplitude after a step response whereas the PI Observer eliminates this error. Requirement 4 concerning the error settling time is met for both designs, hybrid and electric drive.

4.4.4. Controller Design for Test Vehicle

With the IOL a linear system from a new virtual input to the defined system output has been designed. Now a controller regulating the output needs to be developed. As stated in the requirements, a simple controller is desired that can be adjusted easily and comprehensibly by the engineer during experiments. Therefore, a PID controller is selected which is widely used in industrial applications and appreciates an easy understanding and tuning. Since the plant dynamics change for the operation modes hybrid and electric drive, two separate controllers will be designed to ensure satisfactory performance and no tradeoff parameterization.

Since the reference value has an increasing slope like character, there must be two integrals in the open loop of the system to ensure no steady state error due to modeling inaccuracies or disturbances [87]. The controlled plant already shows integral behavior. Therefore, another integral element will be incorporated in the controller. Also P and D elements will be incorporated for good tracking and damping.

The controllers will be designed by shaping the open-loop transfer function in the frequency domain. In the previous section the behavior of the reference system was determined to be an IT1 model. For control design, this model will be the base. It is assumed that if deviations or disturbances are present on the output these will be observed sufficiently fast. This is guaranteed by an observer design which has higher dynamics than the plant itself.

Design Specifications

The depiction of the actuator in the reference model is a simplification that was made due to the applicability of IOL and the properties of the reference model. This simplification is justified as only feedforward signals are processed to make the system input-output behavior linear. For designing a feedback controller though the consideration of time delays is important as significant phase delays at high frequencies occur which lead to a loss in stability margins when closing the loop. An approximation of the limits of the gain crossover frequency for time delayed systems is given by [87]

$$\omega_{gc} \leq \frac{1}{\sum\limits_{i} t_{delay,i}}.$$

(4.28)

Since fast tracking is desired the specifications for the designs therefore are $\omega_{gc} = \frac{1}{\sum_i t_{delay,i}}$ which take on different values for hybrid and electric drive. The second specification is a maximum gain of the sensitivity function to be $M_S \leq 2$ which holds for both designs.

Two controllers are designed by shaping the open-loop transfer function given the two controlled systems described above. The objective of the closed-loop system primarily is to achieve good disturbance rejection. This is motivated by the fact that the system mainly is subject to disturbances acting upon it. Tracking though is not to be neglected as the reference value for the controller is a slope and not a constant value. The main design objectives were introduced in section 4.1 and generalized objectives were described in section 2.3.3. These are now applied to the system defined above.

Although an IOL has been performed the system still possesses another non-linearity which has not been addressed up to this point. It concerns actuator saturation. Due to the restriction to input affine systems this was not incorporated in the IOL and must be dealt with at this point. Methods to deal with this nonlinearity exist [60]. These though cannot be applied here since the non-linearity occurs within the IOL. The influence of saturation though, only occurs with high gains and thus high bandwidths. Therefore, actuator saturation mainly affects the controller design by limiting the bandwidth. This has already been addressed by the specification in equation (4.28) above. Additionally, an anti-windup will be implemented to prevent the integrator from winding up during saturation. The input of the integrator will be set to zero if actuator saturation is reached [119]. For the given problem this is the case when the controller demands more torque than the actuator can provide.

Figure 4.6 shows the Bode plot of the open-loop transfer functions for hybrid and electric drive. For hybrid drive the PID parameters were set so that the gain crossover frequency of 4.8 Hz is reached while still fulfilling the second design specification regarding the maximum sensitivity gain. The sensitivity function of the controller is depicted in figure 4.7 (a). The limit of the sensitivity gain of $M_S \leq 2$ is met. The maximum value is reached at 5.1 Hz. This results in a remaining phase margin of 58° which meets the earlier introduced requirements of the controlled system. Figure 4.7 (b) shows that good tracking is reached for low frequencies without excessive amplification as the maximum value of the complementary sensitivity function is roughly 1.3. If this controller were to be applied to electric drive the system would violate the design specifications

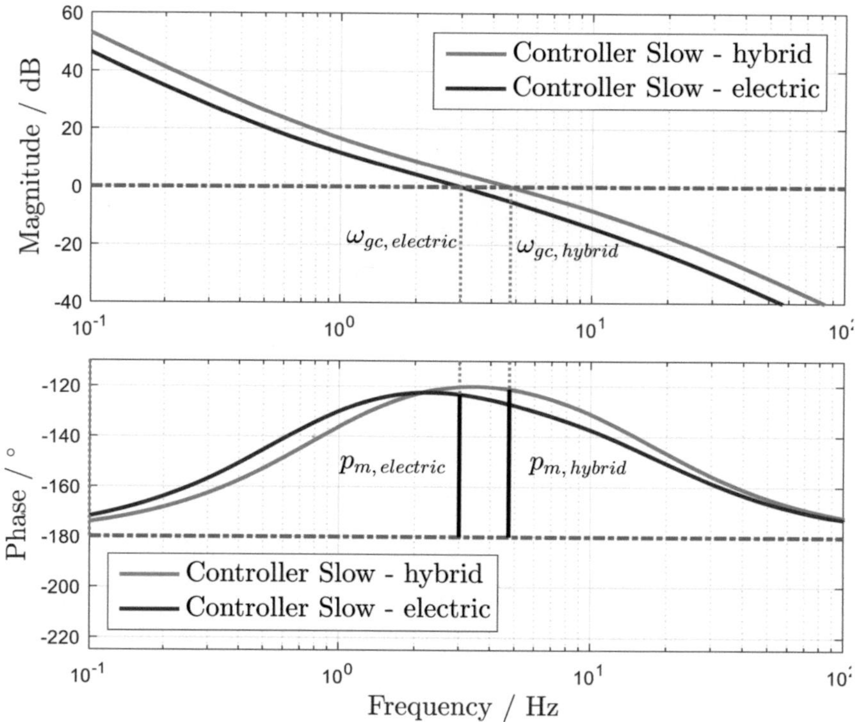

Figure 4.6.: Bode Plot of Open-Loop System for Hybrid Drive and Electric Drive of Controller for Slow Actuators

as the gain crossover frequency would rise above 6 Hz not satisfying equation (4.28). Also, the maximum sensitivity gain would overshoot the limit reaching a value of $M_S > 3$. The controller would not be able to regulate the drivetrain speed as desired.

Due to the reasons explained above a second controller will be designed which will be activated during electric drive. Its values for the PID gains are roughly $\frac{1}{2}$ to $\frac{1}{3}$ compared to those of the controller designed for hybrid drive. This is explained by different drivetrain dynamics as well as actuator dynamics in the reference model. Therefore the controller must be designed more conservatively with a gain crossover of 3 Hz. Satisfying equation (2.16) for the sensitivity gain leaves the open-loop system with a remaining phase margin of $56°$. The complementary sensitivity function for electric drive has a slightly higher maximum value of 1.6 compared to hybrid drive. Applying this controller to hybrid drive would reduce the gain crossover frequency of the open loop to roughly 2.1 Hz. This is a reduction by more than half and would therefore result in a significant loss of performance which is not desirable.

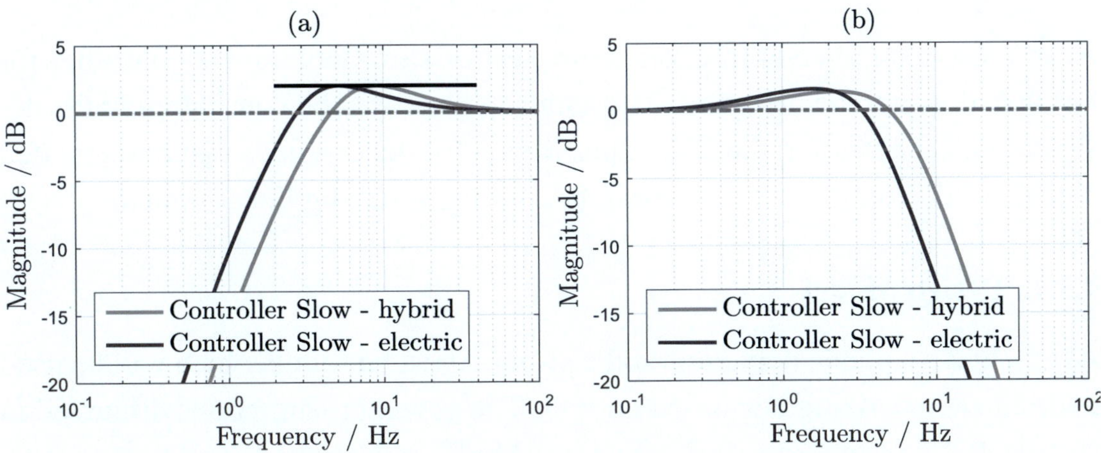

Figure 4.7.: Sensitivity (a) and Complementary Sensitivity Function (b) for Hybrid Drive and Electric Drive of Controller for Slow Actuators

After discussing the controller open-loop system specifications in the frequency domain a brief check of the time domain requirement of settling time is performed. It is directly dependent on the crossover frequency of the system ω_0 and its damping ζ [9]

$$t_{settle,1\%} \approx \frac{4.6}{\zeta \omega_0}. \tag{4.29}$$

Applying this to the controllers designed the settling time of the hybrid drive system is roughly given by

$$t_{settle,hyb} \approx 0.18 \text{ s} \tag{4.30}$$

and of the electric drive system

$$t_{settle,ed} \approx 0.29 \text{ s}. \tag{4.31}$$

Both values are within the requirements.

4.5. Control Design for Fast Actuators

After presenting a novel traction controller for slow actuators, a controller for fast actuators will be designed. This characteristic will be considered in the design model. Thereafter, the IOL is performed for the extended synthesis model. Again, a PID controller will be designed for the resulting linear system.

4.5.1. Design Model

Again, first a suitable depiction of the system must be found which will be used to design a linearizing law based on IOL. The general assumptions discussed in section 4.3 also apply for fast actuators. Model reductions for all components have been addressed. Since for this category actuators are able to damp oscillations an elastic drivetrain depiction is suitable to enhance control performance. The simulation model though incorporates three oscillating masses. For control design purposes this is a high model order. Without losing much depiction accuracy in the desired frequency spectrum a model reduction is performed to achieve a two-mass oscillator. Again, this will be performed with the method of Laschet [77][5]. As proposed by many authors a two-mass oscillator depicting the half-shafts for control design is sufficient [136, 16]. This is also motivated by the dominance of the half-shaft eigenvalues over the two-mass flywheel eigenvalues of factor 3, compare section 3.3.3.

The dynamic equations of the design model are given by

$$\dot{\boldsymbol{x}} = \begin{bmatrix} \dot{x}_1 \\ \dot{x}_2 \\ \dot{x}_3 \\ \dot{x}_4 \\ \dot{x}_5 \end{bmatrix} = \begin{bmatrix} \dfrac{1}{\tau_{red,fast}}(u - x_1) \\ \dfrac{x_3}{i_{tot}} - x_4 \\ \dfrac{1}{J_{mot,red}}\left(x_1 - \dfrac{1}{i_{tot}}T_{hs}\right) \\ \dfrac{1}{J_{whl,red}}\left(T_{hs} - \dfrac{r_{whl}}{i_{tot}}(F_x - F_r)\right) \\ \dfrac{F_x - F_r - F_a}{m_{veh}} \end{bmatrix}, \tag{4.32}$$

$$y = \frac{x_3}{i_{tot}}$$

---

[5]Refer to appendix A.2

with

$$T_{hs} = c_{hs,red}x_2 + d_{hs,red}\left(\frac{x_3}{i_{tot}} - x_4\right), \tag{4.33}$$

which is defined as the half-shaft torque. It is calculated from three states x_2, x_3 and x_4 as well as the drivetrain stiffness and damping coefficients $c_{hs,red}$ and $d_{hs,red}$, respectively. Again, x_1 depicts the motor torque. Introducing an elastic drivetrain to the design model the torsional angle $\Delta\varphi$ of the two masses must be depicted. At half-shaft level it is given by x_2. The rotational speed of the motor shaft is x_3 where the wheel speed is given by x_4. The vehicle speed is depicted by x_5. The definitions of the driving force as well as air and rolling resistance were introduced in section 3.3.2.

The time constant of the resulting actuator system is depicted by $\tau_{red,fast}$. The derivation of the resulting time constant of the system for hybrid and electric drive is described in section 5.2. Again, it will be distinguished between the inertia for hybrid drive $J_{mot,red,hyb}$ and electric drive $J_{mot,red,ed}$. It denotes the motor sided inertia of the two-mass oscillator of the design model. The wheel sided inertia of the design model is denoted by J_{whl} and is independent of the vehicle operating mode.

As system output y, the motor speed scaled to wheel level $\frac{x_3}{i_{tot}}$ is chosen and not the wheel speed x_4 . This can be explained by the speed of information available to the controller. Since the controller will be located on an actuator ECU a signal which is available fast with no latencies is desired. With the given assumption of force closed clutches the signals are identical in steady state conditions allowing this substitution.

4.5.2. Linearization of Plant Dynamics

With the given extended design model (4.7) now IOL is applied to derive a linearizing control law. It addresses the governing nonlinearity of the system, the adhesion of tire and road. First, the output is considered:

$$y = \frac{x_3}{i_{tot}}. \tag{4.34}$$

Now the Lie derivatives of the system output equation will be calculated until a dependency on the input is given. The derivative of y of the analyzed system is

$$\dot{y} = \frac{\dot{x}_3}{i_{tot}}. \tag{4.35}$$

Inserting $\dot{x}_3$ from (4.7) yields

$$\dot{y} = \frac{\frac{1}{J_{mot}}(x_1 - \frac{1}{i_{tot}}T_{hs})}{i_{tot}}, \tag{4.36}$$

With T_{hs} from equation 4.33 the derivative of y is

$$\dot{y} = \frac{1}{J_{mot}\, i_{tot}}\left(x_1 - \frac{1}{i_{tot}}\left(c_{hs,red}x_2 + d_{hs,red}(\frac{x_3}{i_{tot}} - x_4)\right)\right). \tag{4.37}$$

A similarity to the first concept can be observed. The actual motor torque x_1 appears in the first derivative of the system output. The next derivative will show a dependency upon the input u. Here, another reason to choose the motor speed for the linearization becomes apparent. A more direct influence from the input u to the controlled variable is given. Choosing $\frac{x_3}{i_{tot}}$ as system output, a resulting system of second order is achieved. If x_4 had been chosen as system output further derivatives would have to be calculated increasing the system order. This would have made the linearization more complex and dependent on further derivatives which is not desirable due to measurement noise.

Determining the second derivative of y yields

$$\ddot{y} = \frac{1}{J_{mot}\, i_{tot}}\left(\dot{x}_1 - \frac{1}{i_{tot}}\left(c_{hs,red}\dot{x}_2 + d_{hs,red}(\frac{\dot{x}_3}{i_{tot}} - \dot{x}_4)\right)\right)$$

$$= \frac{1}{J_{mot}\, i_{tot}}\left(\left(\frac{1}{\tau_{red,fast}}(u - x_1)\right)\right. \tag{4.38}$$

$$\left. - \frac{1}{i_{tot}}\left(c_{hs,red}(\frac{x_3}{i_{tot}} - x_4) + d_{hs,red}(\frac{\dot{x}_3}{i_{tot}} - \dot{x}_4)\right)\right).$$

The resulting nonlinear controller is dependent on the states x_1, x_3 and x_4 as well as the derivatives $\dot{x}_3$ and $\dot{x}_4$. The derivatives can be calculated exactly by inserting the state equations in (4.32). This however requires exact knowledge of the parameters of the tire-road adhesion and their disturbances. Since this information is not at hand and not estimated reliably a more robust approach is chosen. The derivatives are calculated by using a numerical approximation

$$D(s) = \frac{1}{\tau_D s + 1}, \tag{4.39}$$

where s is the Laplace operator and τ_D being the time constant of the realization filter. It will be chosen small enough to realize a differentiator but high enough to filter noise. Therefore $\tau_D = 0.005$ s is chosen. As the input u is present in the second derivative, the reference model therefore is given by

$$\ddot{y} + a_1\dot{y} + a_0 y = Vw, \tag{4.40}$$

where V, a_0 and a_1 can be chosen freely to achieve a desired behavior. Inserting (4.38) into previous equation and then solving for u yields the new control input for the plant

$$u = x_1 + \tau_{red,fast}\left(J_{mot}\, i_{tot}(Vw - a_1\dot{y} - a_0 y) \right.$$

$$\left. + \frac{1}{i_{tot}}\left(c_{hs,red}(\frac{x_3}{i_{tot}} - x_4) + d_{hs,red}(\frac{\dot{x}_3}{i_{tot}} - \dot{x}_4) \right) \right). \tag{4.41}$$

where w is the new control input. Given the exact linearization, the controllable external dynamics are of second order. This leaves third order internal dynamics that do not have an influence on the system output. However, the system output and its derivatives influence the internal dynamics. An analysis of these internal dynamics for the presented system was published by Reichensdörfer et al. [105]. Global asymptotic stability was proven in this paper by using a parametric Lyapunov function.

4.5.3. Controller Design for Test Vehicle

Before designing the controllers, first the reference model will be discussed briefly. With the given considerations applying an input-output linearization to the extended design model for fast actuators of fifth order again yields a resulting dynamic of second order. This was also achieved for the design model for slow actuators. Therefore, the same motivation and ideas as stated in section 4.4.2.1 concerning the reference model choice are applied. The same integral plant dynamics used earlier will also be applied for the reference model for fast actuators. Only the resulting actuator dynamics change. The derivation of these for hybrid and electric drive are discussed in the following chapter 5.2.3. Note that unlike for slow actuators the dynamics for the fast actuators remain similar for hybrid and electric drive.

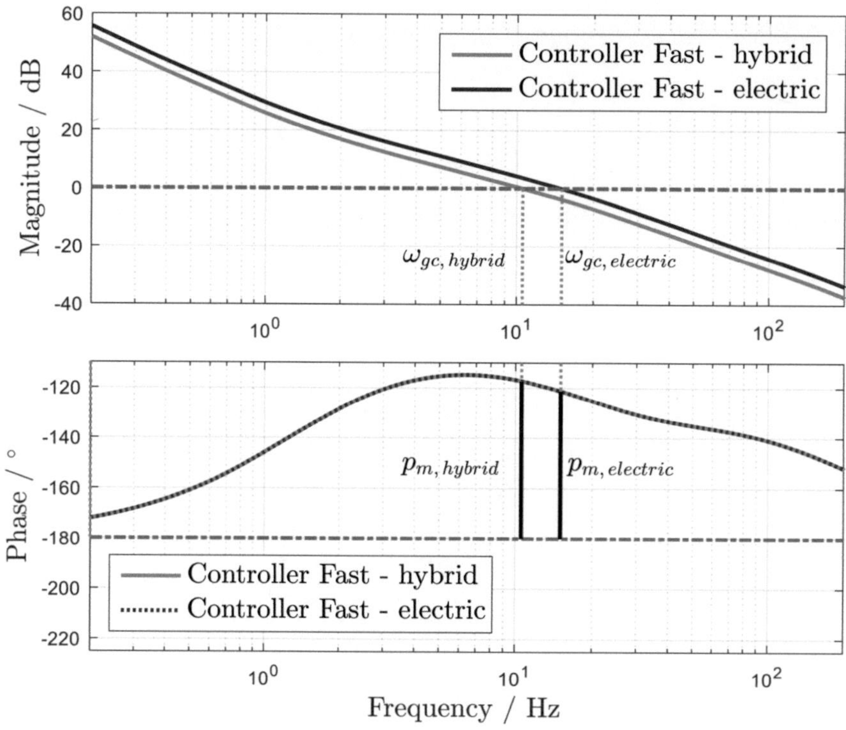

Figure 4.8.: Bode Plot of Open-Loop System for Hybrid Drive and Electric Drive of Controller for Fast Actuators

Due to similar and especially fast actuator dynamics, only one controller will be designed. This satisfies the requirement of "Convenient Parameter Handling" which can be assessed to the group "Development", refer to section 4.1. Parameterizing only one controller in experiments is faster and reduces implementation complexity of the controller. Again, a PID controller is chosen as it is a simple, widely known, and intuitive structure which can be adjusted easily in experiments to meet the desired behavior in all driving maneuvers. It will be designed by shaping the open-loop transfer function in the frequency domain. The same design specifications introduced in section 4.4.4 hold for fast actuators. For slow actuators, the crossover frequency was determined by the sum of time delays within the feedback loop. For fast actuators though, this target is not feasible due to another nonlinearity which has not been addressed yet. It concerns the rate limitation of electric machines. Simulations have shown that a good tradeoff between the consideration of rate limitation and performance lies in a gain crossover frequency of roughly 15 Hz which will be set for electric drive. Due to the different plant dynamics of the reference model the controller will be designed using the electric drive parameters. If this controller fulfills the specifications these will also be met when applying the same controller for hybrid drive.

The open-loop Bode plots are shown in figure 4.8. The controller for electric drive leaves a phase margin of 59° while still limiting the maximum sensitivity to $M_S \leq 2$, see figure 4.9. This controller applied for hybrid drive results in a gain crossover frequency of 10.5 Hz with a maximum sensitivity gain of $M_S = 1.6$. A phase margin of 62° is left offering good robustness.

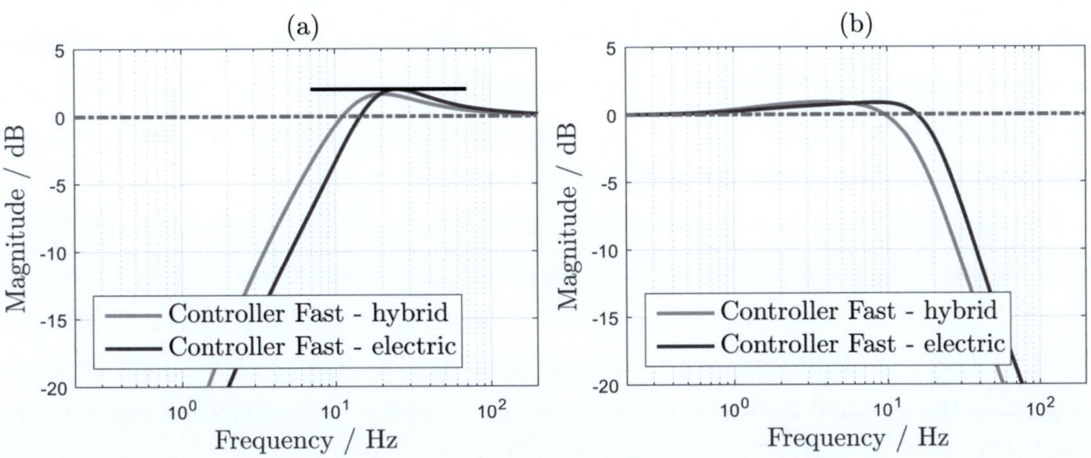

Figure 4.9.: Sensitivity (a) and Complementary Sensitivity Function (b) for Hybrid Drive and Electric Drive of Controller for Fast Actuators

Applying equation (4.29) to the fast controllers the settling time of the hybrid drive system is roughly given by

$$t_{settle,hyb} \approx 0.09 \text{ s} \tag{4.42}$$

and of the electric drive system by

$$t_{settle,ed} \approx 0.08 \text{ s}. \tag{4.43}$$

Both values are within the requirements.

4.6. Simulation and Comparison of the Designs

After proposing two control concepts and designing controllers for each based on the parameters of the investigated vehicle, these will be compared in the following. This is done in a simulation environment. As control allocation has not been discussed at this point only electric drive will be chosen for the comparison. The integration within the ECU network is motivated in chapter 5.3 and applied to this simulation. A complete validation of the concepts for all drive modes and various maneuvers will be shown in chapter 6. The present section only serves as a verification of the designed controllers before continuing with control allocation design.

For the comparison a critical maneuver is chosen, a step in friction coefficient from high to low during high acceleration[6]. The results for the slow actuator concept are depicted in figure 4.10 (a) whereas the results for the fast actuator concept are depicted in (b). The top shows rotational drivetrain speeds and the bottom desired and actual torques of ICE and EM.

The concept for slow actuators shows a significantly higher overshoot of factor 2 of the EM rotational speed. This is the result of the lower actuator dynamics and therefore, lower open-loop gain crossover frequency of the controller design. High oscillations of the driven axle occur. These though are not reflected in the control action due to the PI observer which observes the virtual, rigid drivetrain speed. The first oscillation peak of the rear axle is still present for the fast controller but the following oscillations are damped effectively. The reason for this is the higher gain crossover frequency of the open-loop system. More important though is the depiction of the elastic drivetrain in the design model. This IOL law incorporates the following term

$$-\frac{c_{hs,red}}{i_{tot}}\left(\frac{x_3}{i_{tot}} - x_4\right). \tag{4.44}$$

This term was suggested by many authors to damp driveline oscillations [119, 136]. Additionally, the author showed that this damping term enhances acceleration potential during TC of a high-performance ICE vehicle [141].

Another difference lies in the resulting actuator dynamics. For the slow actuator concept a significant time delay is observed which reflects the bus communication time. Rate limitation is not an issue due to the conservative pa-

---

[6]Refer to section 6.1 for details on driving maneuvers.

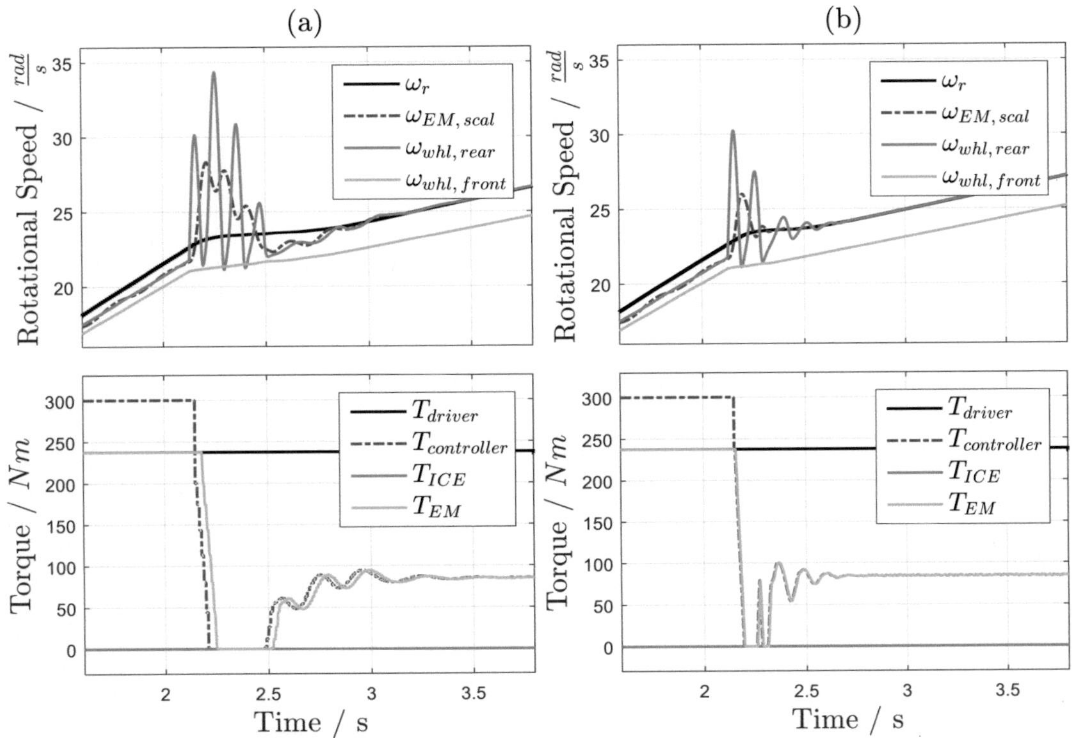

Figure 4.10.: Simulation of Controller Designs for Electric Drive of Test Vehicle Model for Step in Friction Coefficient from High to Low - Concept for Slow Actuators (a) and Concept for Fast Actuators (b)

rameterization. The control action for fast actuators though is restricted by rate limitation from start of the control at 2.15 s until 2.3 s. Due to the limitation of the open-loop bandwidth during design this does has no significant effect on the control task.

As both simulations have the same starting condition the final speed of the undriven wheels are an indicator for the performance of the controller. It shows the degree of utilization of the maximum tire-road adhesion potential. The simulation shows that the final speed is higher when regulated with the controller for fast actuators. At this point only a qualitative comparison is made. A quantitative comparison will follow in section 6.2 with an experimental study.

5

Control Allocation and Implementation

In parallel HEVs there are two ECUs that operate the torque generating actuators of the powertrain. This offers a degree of freedom to implement a speed-based slip controller. To motivate distinctive designs of implementation, preliminary considerations must be made. These consist of two steps. First, individual actuator dynamics will be analyzed when controlled from a given ECU. This will be based on the identified model from chapter 3. The results are used to motivate the applied control allocation method. Additionally, the motivation of the application of the controller designs is presented. The study results in two distinct design options of the TC system. The chapter closes with a comparison of both proposed designs based on a simulation. The following sections are based on the contribution [140].

5.1. Allocation Concept for CECU

5.1.1. Individual Actuator Dynamics

Given the system description of ECU and actuators, as well as the parameterization, the dynamics of the individual actuators when controlled from the CECU will be displayed. For this, a step in desired torque is chosen as it excites all frequencies and shows occurring time delays well. The step responses are shown in figure 5.1. Similar dynamics can be observed. Both actuators show a pronounced time delay in their response to the desired value.

The step responses differ in time delay and time constant. To compare both step responses, a single, decisive parameter is desired. Lunze [87] and Kuhn [75] described a method to reduce the order of a system to a first order low pass:

$$G(s) = \frac{(\tau_{D,1}s + 1) \cdots (\tau_{D,n}s + 1)}{(\tau_1 s + 1) \cdots (\tau_m s + 1)} e^{-t_{delay,1}s} \cdots e^{-t_{delay,o}s}. \tag{5.1}$$

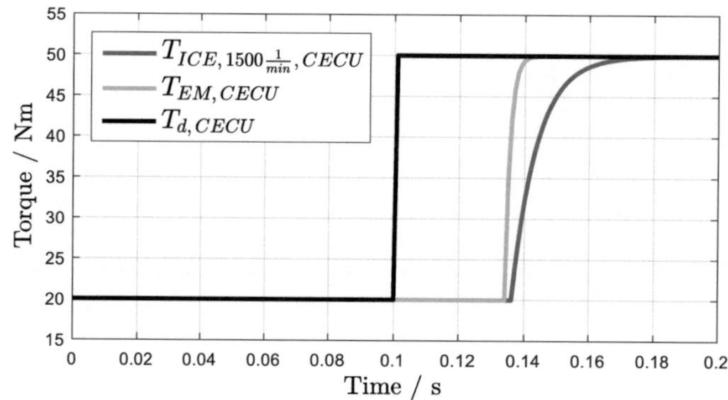

Figure 5.1.: Step Response of ICE and EM when controlled from CECU

For this, all retarding time constants τ_i and time delays t_{delay} of the system are added and all differential time constants $\tau_{D,j}$ are subtracted:

$$\tau_{red} = \sum_{i=1}^{n} \tau_i - \sum_{j=1}^{m} \tau_{Dj} + \sum_{k=1}^{o} t_{delay,k} \,. \tag{5.2}$$

Applying this method to the system yields reduced time constants of

$$\tau_{red,ICE,CECU} = 0.043 \text{ s}, \tag{5.3}$$

$$\tau_{red,EM,CECU} = 0.035 \text{ s}, \tag{5.4}$$

for the ICE and EM, respectively. As they lie in the same magnitude order, daisy chain allocation can be applied, compare 2.4.1. It is commonly used in parallel HEVs [90] and is active during hybrid drive in stable driving maneuvers. It is applied in state-of-the-art direct slip TC. However, it has not been applied for speed-based TC. It will therefore find novel usage as allocator concept for speed-based TC. It can be effectively used for this implementation as the above analysis shows.

5.1.2. Implementation

The daisy chain method only offers one design option, the prioritization of the actuators. For the present case, the ICE is chosen as it is the most powerful actuator providing enough drive torque for an extended period of time.

A simplified depiction of the allocation concept is shown in figure 5.2. First, the load point shifting torque T_{LPS} is added to the total desired torque T_d to satisfy the energetic operation strategy of the high voltage system. The sum is limited by the primary actuator constraints. The resulting desired torque $T_{d,ICE}$ is then limited and transferred to the ICE. The difference of the desired torque for the ICE and the total desired torque is then transferred to the EM as its desired torque $T_{d,EM}$.

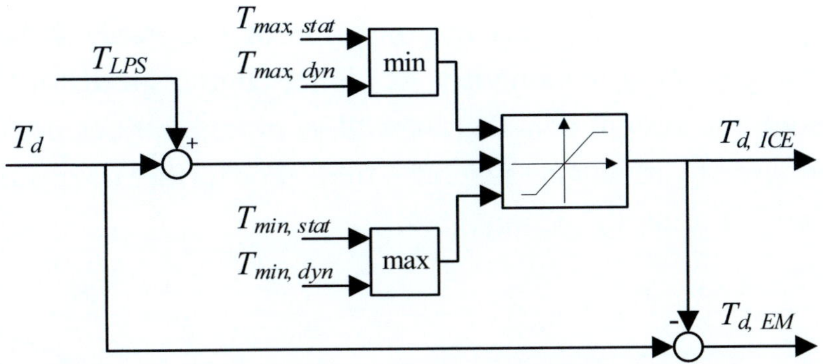

Figure 5.2.: Control Allocation Concept CECU

The limits for $T_{d,ICE}$ are divided into steady state and dynamic limitations. Each have lower and upper boundaries. The steady state limitations, $T_{max,stat}$ and $T_{min,stat}$, are dependent on the operating point, in this case the crankshaft speed. The dynamic limitations, $T_{max,dyn}$ and $T_{min,dyn}$, are additionally dependent on the actual torque and cylinder fill. Here, the limitations are defined by the dynamics of the air path which divide into faster naturally aspirated and slower turbo dynamics. At this point no torque reduction via ignition timings is tolerated due to limitations of exhaust emissions. It is assumed that there is access to the information on the available dynamics of the ICE to a high degree of accuracy. These are determined by a complex model in the ECU software, compare section 2.2.

This allocation concept must also consider pure electric drive. For this, simply the minimum and maximum torque limitations of the ICE are set to zero. Now the desired torque for the ICE remains zero and all requested torque is allocated to the EM. The control logic is handled by a state machine which will not be described at this point since this work does not focus on state transitions between hybrid and electric drive.

5.1.3. Dynamic Properties of Allocation and Actuators

For control design purposes the analysis of the combined system is important. First, the system for hybrid drive is analyzed. Figure 5.3 (a) shows the Bode plot of the resulting system from virtual control input (desired torque) to combined actuator output[1] for two distinct engine speeds at $1500\,\frac{1}{min}$ and $5000\,\frac{1}{min}$. It includes the daisy chain distribution as well as actuator dynamics. Also, approximations of the system as first order low pass filters are plotted as dashed lines. These approximations were utilized in section 4.4 for controller design. At higher speeds, a faster system response can be observed. This can be explained by the physical functionality. At higher engine speeds of the ICE, the airflow through the system is higher as well as more ignitions per time period occur which directly affects the response time. For further details on the functionality of an ICE refer to section 3.3.4.1.

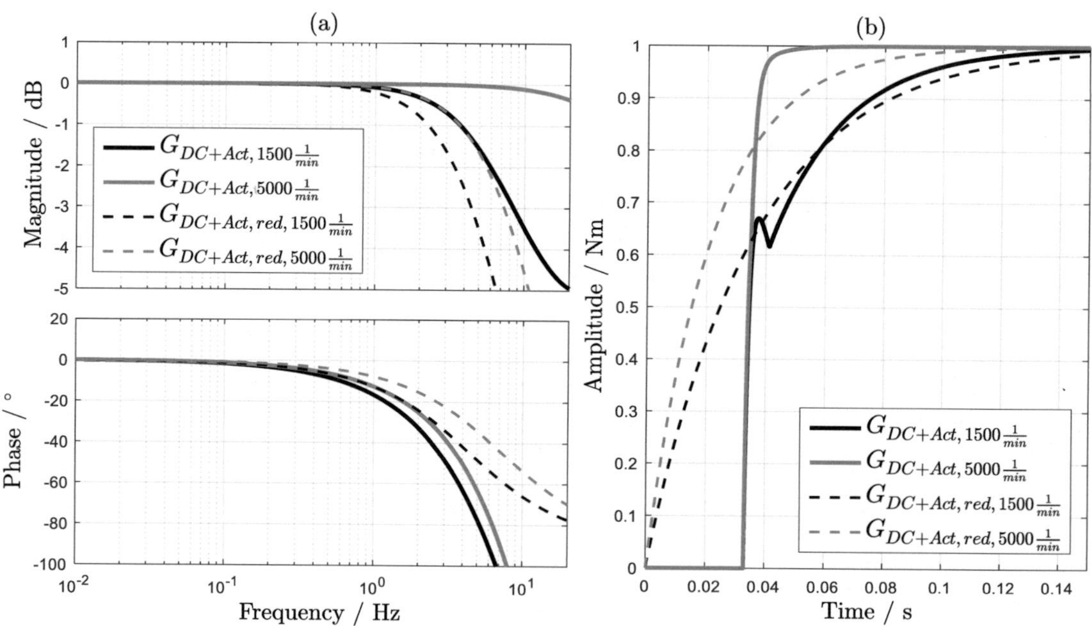

Figure 5.3.: Dynamic Properties of Daisy Chain and Actuators - Bode Diagram (a) and Step Response (b)

Step responses of the combined system of daisy chain and actuators as well as approximations of it as first order lowpass (dashed lines) are depicted in figure 5.3 (b). The approximations are determined by a parameter optimization varying the time constant of the reference system which is described in [87] to minimize the value of the quality function. The quality function evaluates the area between the reduced order system step response and the actual system response. This method is chosen since it yields a rather good approximation with the actual system response in the time domain. In the frequency domain

---

[1]Combined actual torque of EM and ICE

a tradeoff between phase and magnitude can be observed. A more accurate approximation of time delays in the frequency domain is given by the Padé approximation [122]. This though is not applied at this point since it yields a transfer function of high order which is contrary to the objective of system order reduction. The actuator system for electric drive will be approximated by the method of Kuhn [75].

5.2. Allocation Concept for EECU

5.2.1. Individual Actuator Dynamics

Again, the dynamics of the individual actuators will be analyzed; now, controlled from the EECU. The step responses are shown in figure 5.4. As opposed to the results in 5.1.1, in this case the dynamics of the step responses differ, especially regarding the time delay.

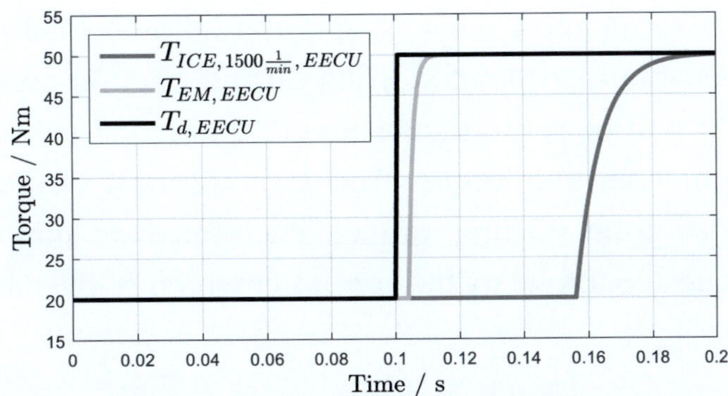

Figure 5.4.: Step Response of ICE and EM when controlled from CECU

Performing the method of Kuhn yields

$$\tau_{red,ICE,EECU} = 0.063 \text{ s}, \tag{5.5}$$

$$\tau_{red,EM,EECU} = 0.005 \text{ s}, \tag{5.6}$$

for ICE and EM time constant, respectively. These differ in one magnitude order. Given this, an allocation concept that can deal with this property is preferred. A promising solution for this problem is the dynamic control allocation as the literature review, and fundamentals in sections 1.2 and 2.4, respectively show. This though has not yet been applied to TC problems. Furthermore, the parameterization of this CA method has not been addressed subjectively in known publications. The following section will therefore propose a method to this open research field.

5.2.1.1. Method for Parameterization

When designing the DCA there are several parameters to determine. Considering equations (2.21) three matrices W_1, W_2 and W_v appear that must be calibrated. Additionally, the steady state vector offers a degree of freedom which must be addressed and finally the control effectiveness matrix B must be parameterized.

The dynamic control allocation has three inputs. One is the virtual control effort v, the other two are the desired steady state values of both actuators u_s, see equation (5.7). Usually the steady state torques of a hybrid vehicle are provided by the ICE. Given the same reasoning as for the daisy chain approach in Design A this will be adopted for the DCA. Therefore, the steady state input of the ICE will be set to the virtual control input whereas the steady state input of the EM will be set to zero. This assumption holds if load point shifting is deactivated. To enable load point shifting during traction control the desired torque for the ICE will be set to the desired torque requested by the controller adding the load point shifting torque. The desired torque of the EM then is set to the negative load point shifting torque. Therefore, the sum of both torques remains v, the requested torque by the controller which is given by the following equation

$$u_s = \begin{bmatrix} u_{s,1} \\ u_{s,2} \end{bmatrix} = \begin{bmatrix} u_{s,vm} \\ u_{s,em} \end{bmatrix} = \begin{bmatrix} v + T_{LPS} \\ -T_{LPS} \end{bmatrix}.$$ (5.7)

Second, the control effectiveness matrix will be addressed. This matrix defines which effect each actuator position has on the desired system input. In the present case the torque ratio each individual actuator applies to the gearbox input shaft is of interest. Since both actuators are connected without further transmission ratios both entries of B are set to 1.

Matrix W_v is used to penalize the virtual control error. In the present case no control error will occur. This is guaranteed by respecting the static and dynamic limits of the system in the controller as its output is limited. For this reason, W_v will be parameterized arbitrarily with the value 1. This leaves only W_1 and W_2 to be determined which will be discussed in the following.

The proposed approach for parameterizing the DCA is based on an optimization. The weighting matrices are determined by running a simulation model of the investigated process for a distinct maneuver. Notice that the parameterization can only be performed for a known implementation which will be described in section 5.3. Design B will be assumed for this parameterization as it makes use of the DCA. The approach consists of the following steps

1. Definition of Quality Function

2. Choice of Maneuver

3. Determining Weighting Parameters of the Quality Function

4. Optimizing Weighting Matrices

Quality Function

For the optimization process a quality function is required. It takes the two main objectives into account, meeting steady state objectives and minimizing the offset of control effort. This function is given by

$$J = W_1 \Delta t + W_2 A \tag{5.8}$$

with A being the area between virtual and actual control effort and Δt the settling time of the actuators to their steady state values. Since both measures have different units and absolute values the factors W_1 and W_2 are introduced. With these, both measures can be weighted equally leading to a balanced evaluation.

The settling time is determined by

$$\Delta t = t_{steady state, 1\%} - t_{start} \, . \tag{5.9}$$

It depicts the difference between the absolute time when the steady state values are reached $t_{steady state}$ and the start time t_{start} of the maneuver. The start time is the absolute time when the maneuver begins. To determine the steady state time a band around the steady state value of $+-1\,\%$ of the actuators maximum steady state value is generated. The steady state time is set as soon as the actual control effort of the actuator falls into this band.

The area A of the quality function is determined by

$$A = \int_{t_{start}}^{t_{start}+\Delta t} |v - \boldsymbol{B}\boldsymbol{y}|\, dt \tag{5.10}$$

where y is the actual control effort produced by the individual actuators. It is defined by

$$y = \boldsymbol{G}_{actuators}\boldsymbol{u}$$

where $\boldsymbol{G}_{actuators}$ is the set of transfer functions of the actuators.

Driving Maneuver

Since the control allocation is part of a closed-loop control system, the measures are determined for a distinct maneuver, usually defined by a step in the reference value or a step in the disturbance influence. For the present case, the latter is chosen which is a step in the friction coefficient from high to low since it is the most critical maneuver in TC and poses the highest stress on the controlled system. The assumption here is that an optimally adjusted control allocation for the most stressful maneuver also provides good allocation performance during less critical maneuvers. Details on the maneuver are described in section 6.1.

The simulation requires a parameterized controller providing the desired control effort. For this the designed controller in section 4.5 will be used.

Weighting Parameters

Before beginning the optimization process the weighting parameters a_1 and a_2 must be determined. For this the chosen maneuver will be simulated without the DCA and only with the fastest actuator. The fastest actuator is chosen since it possesses the best properties to regulate the system. From this initial simulation, the measures Δt and A of the quality function are determined and will be named $\Delta t_{initial}$ and $A_{initial}$. The control area will be determined by equation (5.10). The initial settling time will be determined by the time it takes for the controller to compensate the disturbance. It is defined as the time from controller activation until the control error falls into a band of $+-1\%$ around the desired value. The measure is defined this way so that the steady state values of the actuators are reached as soon as the controller has regulated the system output to its desired value.

Using the measures that resulted from the initial ideal simulation the weighting factors can be determined. Since an equal weight of time and control area is strived for the weighting factors are set to

$$W_1 = \frac{W_3}{\Delta t_{initial}}$$
$$W_2 = \frac{1}{A_{initial}}, \tag{5.11}$$

with $W_3 = 1$, an additional factor to prioritize either measure. By changing W_3 in the above equation, an unequal weighting of both measures is made possible. A higher value of W_3 for instance prioritizes the settling time.

Optimization

Having defined the quality function (5.8) and the parameterization of the weighting factors (5.11) the optimization of the weighting matrices can be performed. This will again be done by simulating the chosen maneuver. For this step though all actuators will be activated. For the present problem hybrid drive will be set.

The chosen optimization algorithm is the simplex method introduced in section 3.3.1. Again, the properties of this algorithm are suitable for the given optimization problem. Only the values of the weighting matrices W_1 and W_2 need to be determined. These are diagonally occupied quadratic matrices of the size 2×2. Therefore, four entries exist which will be optimized. Additionally, the starting values are known to a reasonable degree of accuracy to be close to the optimal values.

Results

The proposed method yields on optimal parameterization of the DCA for the chosen maneuver. Given equation (2.24) these results are transformed to a set of linear filters of order two. Now the dynamic properties of both filters can be analyzed. The Bode diagrams of these are depicted in figure 5.5. The maximum frequency of the filters lies at 500 Hz due to the sampling theorem by Nyquist and Shannon [86]

$$\tau_{sample} < \frac{1}{2 f_{max}}.$$

It states that the maximum observable frequency f_{max} of a time discrete system is smaller than half of the inverse of the sample time τ_{sample}.

The first filter (a) has a low pass character for low to medium frequencies with a cut off frequency of about 0.7 Hz and addresses the ICE. This can also be seen in the phase plot where the phase starts at $0°$ for low frequencies and lowering to almost $-90°$ for medium frequencies. From medium to high frequencies a phase lifting behavior is observed which also reduces the magnitude loss to $0\frac{\text{dB}}{\text{decade}}$. The second filter (b) has a high pass character which also has a cut off frequency of approximately 0.7 Hz. The phase starts at $90°$ passing $45°$ at the cut off frequency and then reaching $0°$ for high frequencies. This filter addresses the EM.

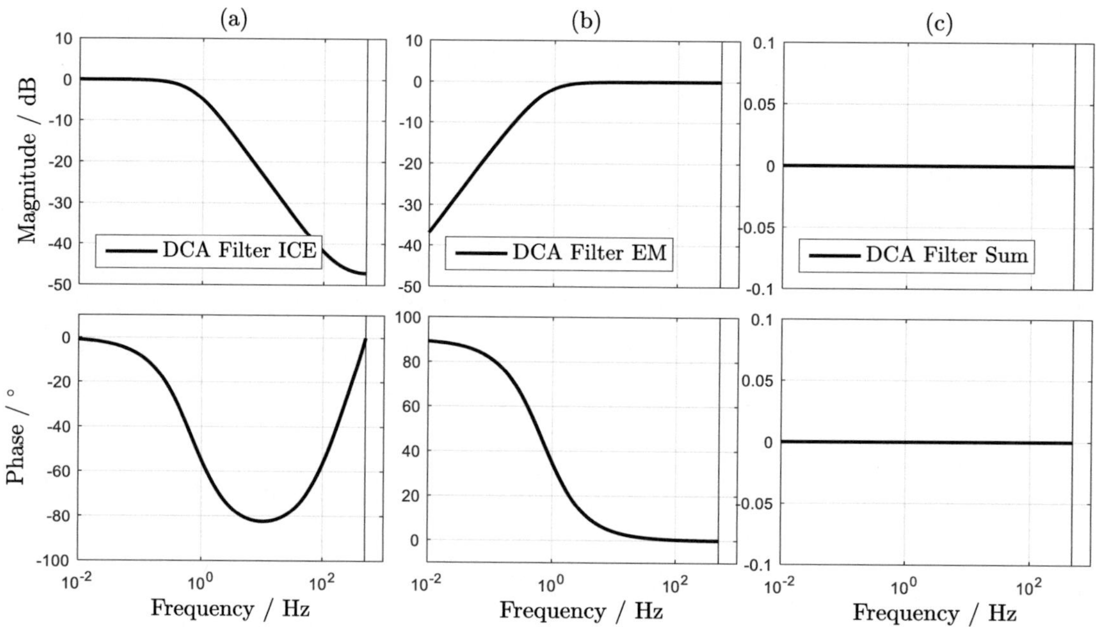

Figure 5.5.: Bode Diagram - Dynamic Properties of DCA Filters

The exact parameters of the optimization results can be found in appendix A.4.

Discussion

An optimal set of filters for the task of control allocation for the chosen maneuver have been created. Before continuing to the implementation of the parameterized DCA, the filter properties will be analyzed further. Equation (1.2) reformulated yields

$$\frac{v}{Bu} = 1, \tag{5.12}$$

For this equation to hold, both filters must be complementary so that no information is lost throughout the entire frequency spectrum. Therefore, magnitude and phase of both filters combined must equal 0. Therefore,

$$|G_{CA,ICE}(j\omega) + G_{EM}(j\omega)| = 0,$$
$$\arg(G_{CA,ICE}(j\omega) + G_{EM}(j\omega)) = 0. \tag{5.13}$$

must hold. The Bode diagram of the sum of both transfer functions is depicted in figure 5.5 (c) proving the assupmtion[2]. Magnitude and phase remain at 0 for the entire frequency spectrum. Notice that this remains true for the task of distribution of desired torque only. This depiction does not incorporate the conversion to actual torque. The dynamic properties of the combined system of control allocation and actuators will be discussed in section 5.2.3.

5.2.2. Implementation

The implementation of an optimization process with uncertain number of loops is not desired for industrial vehicles. This can be explained by the uncertain computational intensity and non-deterministic behavior of this approach. For this reason, the explicit solution is made use of which was introduced in equation (2.23) and depicted in figure 2.9. The ICE torque demand is filtered by G_{ICE}. Additionally, a saturation for the ICE torque will be included. This is important due to its pronounced nonlinearity in its dynamics caused by turbo charging. Torque that cannot be provided by the ICE is shifted to the EM. Instead of implementing two filters as the direct solution proposes the fact that these filters are complementary is utilized. The torque for the EM is determined by $1 - G_{ICE}$.

Simulations have shown a reduction of computational intensity by a factor of 50 when using discrete filters instead of the optimization process. In simulations the exact behavior of the optimization process was achieved by the simple solution of one discrete filter and torque shifting during actuator saturation. Similar results concerning the computational intensity were published in [113]. Here, regenerative braking using hydraulic brakes and an EM was studied.

This idea extends the explicit solution by incorporating the actuator limitations. This modified explicit solution now performs in the same way as the DCA with the optimization process. It has the character of the daisy chain method combined with the explicit solution of DCA. Figure5.6 shows the schematic

---

[2]This can also be determined by examining the discrete transfer functions of the filters. Due to the same absolute values but with opposite sign all factors cancel each other out.

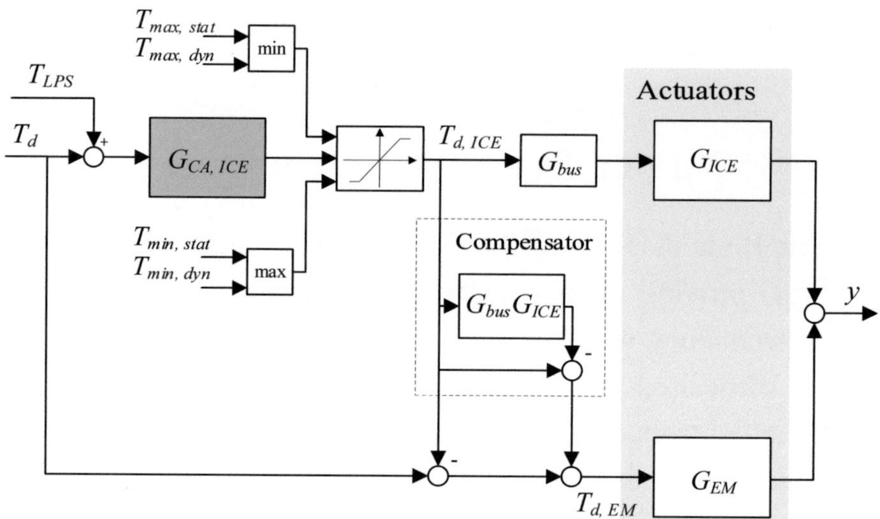

Figure 5.6.: Integration of DCA functionality to EECU

of implementation of the DCA approach to the EECU of the test vehicle. It includes a compensation term which will be described in the next section.

5.2.3. Dynamic Properties of Allocation and Actuators

The previous section described the allocation algorithm and its dynamic properties. The system inputs were the virtual control input and the output the desired values for the actuators. Now the full transfer behavior from virtual control input to actuator output will be discussed. The block diagram of the system is depicted in figure 5.6 with G_{ICE} being the transfer function of the ICE including the bus communication time and G_{EM} being the transfer function of the EM introduced in chapter 3.

Plotting the transfer function from v to y reveals a bumpy response in the frequency domain see figure 5.7. It can be explained by the pronounced time delays during communication. This is an undesired behavior as it lies in the core frequency spectrum of the controller.

To raise performance of the CA a compensator is introduced. It is depicted in figure 5.6. This compensator makes use of the same transfer function as the ICE including the signal delays due to bus communication. The difference of the desired value communicated to the combustion engine and its actual value from the model is added to the desired signal of the EM. The benefit of this compensator is depicted in figure 5.7. The Bode diagram of the transfer functions with and without the compensating term are depicted for an engine speed

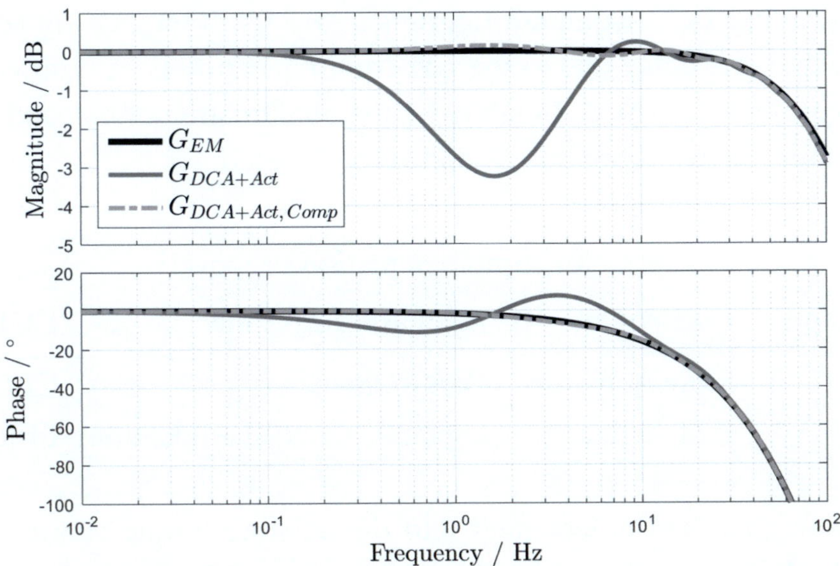

Figure 5.7.: Bode Diagram - Dynamic Properties from Virtual Control Input to Actuator Output

of $1500\ \frac{1}{\text{min}}$. Without the compensation the magnitude decreases by more than 3 dB in the dominant frequency spectrum of 1 to 10 Hz. Also shifts in the phase can be observed which is undesirable. These unwanted bumps are evened out by the compensator making the new system now roughly behave like the EM. As a reference the Bode diagram of the EM response is also depicted in the plot.

Since the combined system of daisy chain and actuators behave like the EM itself the method of Kuhn will be used for model reduction to a first order low pass. This makes it applicable to the method of input-output linearization in chapter 4.4.1 as only actuators behaving like a first order low pass are considered.

5.3. Implementation of Traction Control System

Parallel HEVs offer a degree of freedom for the implementation of speed-based slip control. In the previous sections and chapters, two controllers for speed-based TC, and two control allocation methods were proposed. Based on these results, two implementation proposals will be made. They will be motivated by the existing control structures in industrial vehicles. For both implementation proposals the reference value calculation will be located on the driving dynamics ECU (DECU). The reference wheel speed is communicated by bus to the controlling ECU, compare figure 4.2. As the reference signal has low dynamics, the time delay due to the bus communication is insignificant to the overall control performance. In table 5.1 the combination of CA and designed controllers for the two proposals, **A** and **B**, is depicted:

Table 5.1.: Implementation Proposal

Design	Controller	Allocation Method	ECU
A	Slow Actuators	Daisy Chain	CECU
B	Fast Actuators	Dynamic Control Allocation	EECU

These now must be implemented into the existing torque control structure of parallel HEVs which was introduced in section 2.2. Figure 5.8 shows the implementation of Design A within level 1 driving functions. The slow traction controller C_{slow} will be added to the torque control path limiting the driver torque. This yields the coordinated torque T_{coord}. It is distributed to ICE and EM using daisy chain control allocation DC which provides the desired torques for ICE $T_{d,ICE,C}$ and EM $T_{d,EM,C}$. These are then communicated to the LLCs of the individual actuators. The various arrows exiting the ECU and entering the actuator depict the LLC output. In Design A, the CECU will act as the master ECU at all times.

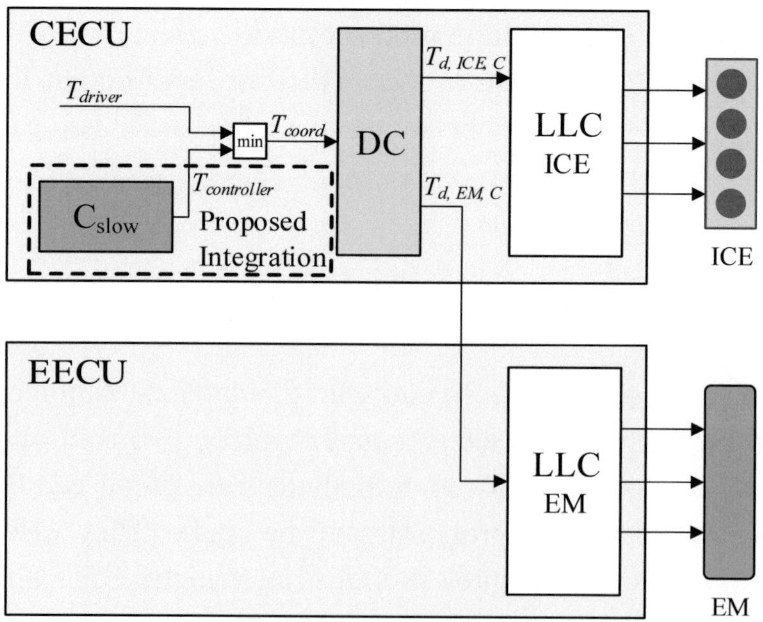

Figure 5.8.: Integration of Traction Control Concept on CECU, Design A

Notice that the signals crossing ECU boundaries are communicated by bus and are therefore delayed. Figure 5.8 focusses on a high-level functional depiction of TC. Further ECU content and additional ECUs purposely will not be displayed. This also holds for figure 5.9.

Design A only requires model changes in CECU software. There are no changes to bus communication as communication of desired EM torque to the EECU already exists. Also, no changes in the control allocation is required as HEVs are already equipped with a CA which matches the requirements set by the controller. Therefore, the software change regarding Design A only concerns the implementation of the controller. The designed controller adds roughly 4 KB of code with a calibration data size of 200 Byte.

The implementation of Design B is depicted in figure 5.9. The fast traction controller C_{fast} is located on the EECU. The driver torque is limited, and the resulting torque is distributed by dynamic control allocation *DCA*. Its desired torque for the EM $T_{d,EM,E}$ limits the desired torque for the EM communicated from the CECU and is passed on to the low-level torque control of the EM *LLCEM*. The desired torque for the ICE $T_{d,ICE,E}$ is communicated to the CECU and limits the maximum torque of the ICE. It is incorporated into the torque allocation and therefore affects its outputs. The designed controller including CA also adds roughly 4 KB of code with a calibration data size of 350 Byte[3] to the EECU. The computational intensity on the other hand rises tenfold due to the tenfold higher sampling frequency.

For Design B, modifications in the existing communication of drivetrain ECUs must be made. Depending on the initial communication, up to two additional signals must be added to bus communication. If not received by the EECU already, the driver torque needs to be added to the bus communication, preferably with a 10 ms sampling. This is necessary as it contains high gradients during tip-in and tip-out maneuvers. Also, a resolution of 12 bit will be required for exact comparison with the controller torque. As this signal is necessary for the activation of the traction controller an additional qualifier is desired. It requires the same sampling rate and a resolution of 4 bits. An even number is desired as qualifiers carry a set of different stati: "initializing", "signal valid", "signal invalid" etc. With this information the behavior of level 1 functions can be determined, and level 2 functions can monitor the behavior.

---

[3]The different data types of 32 Bit for functions on the EECU compared to 16 Bit on the CECU results in a higher demand. The number of application parameters remains roughly the same. This is explained by not using an observer for the drivetrain speed but therefore incorporating DCA

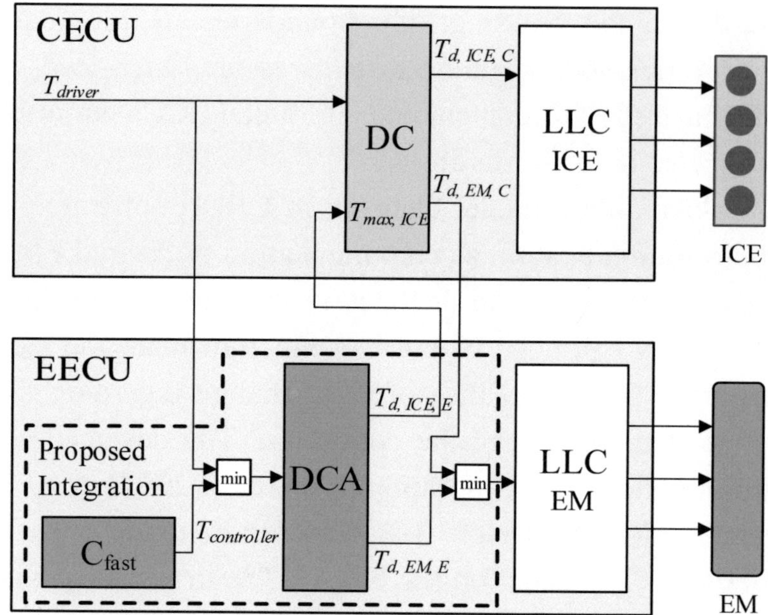

Figure 5.9.: Integration of Traction Control Concept on EECU, Design B

The second additional signal is the limitation of the ICE torque which is transmitted from EECU to CECU. Also, a sampling rate of 10 ms is desired with a resolution of 12 bit. Since powertrain bus systems operate on CAN which has a transfer rate of up to 1 Mbit/s both additional signals add 0.3 % of data rate to the system. A major distinction between Design A and B is the switching of master ECU which controls the torque demand from CECU during regular drive to EECU during TC. For this reason, the additional signals on the bus are required.

Additionally, the implementation of Design B not only affects functionality in EECU where controller and CA are located but also on CECU. Additional signals must be routed to the COM-Adapter and must be modeled in the function itself. The highest impact though remains on EECU.

Both designs require a change in communication from DECU to either powertrain ECU depending on the design. Although the maximum wheel speed has low dynamics, a rate transfer of 10 ms and 16 bit word length is required. This can be explained by its slope like character and a precise desired value necessary for the overall TC performance. As the DECU is connected via Flexray to the powertrain ECUs and offer up to 10 Mbit/s, the signal plus qualifier add 0.02 % of data rate.

To sum up, the low computational intensity and stack demand of both proposals ensures integrability to state-of-the-art ECUs. This satisfies a major requirement introduced in section 4.1. Also, an integration to the communication system via bus is ensured due to low bandwidths of just two to three signals. Table 5.2 summarizes the results discussed in this section.

Table 5.2.: Required Resources of Designs

	Design A	Design B
ECU	CECU	EECU
Sample Time	10 ms	1 ms
Rom Code	4 KB	4 KB
Ram Data	0.2 KB	0.35 KB
Powertrain Bus Increase	0 %	0.3 %

5.4. Simulation and Comparison of the Designs

After proposing two implementation concepts of the designed controllers and control allocation concepts these will be compared in the following. Again, a simulation environment is chosen as a first verification step.

For the comparison again a step in friction coefficient is chosen. Figure 5.10 depicts the results of the CA design for Design A for slow actuators (a) and Design B for fast actuators (b). The top shows rotational drivetrain speeds and the bottom the desired torques of driver, controller as well as ICE and EM. The more incremental desired torques in figure (a) is due to the lower sampling frequency of the computing ECU.

The controller torque of both designs have a similar progression with the exception that Design A is smoother which is traced back to the control algorithm and its parameterization. Concerning control allocation there is a major difference. Design A shifts high dynamic parts of the control action to the EM but also to the ICE if the dynamic limitation permits. This results in an undershoot of 70 Nm below the steady state controller torque at 2.4 s for the ICE. Directly thereafter an overshoot can be observed. The EM receives the delta torque letting it deviate from the desired torque defined by load point shifting. High gradients are shifted to the EM although resulting in high phase delays during communication since shifting to the ICE would be worse. This low dynamic resulting actuator system is reason for the conservative controller calibration.

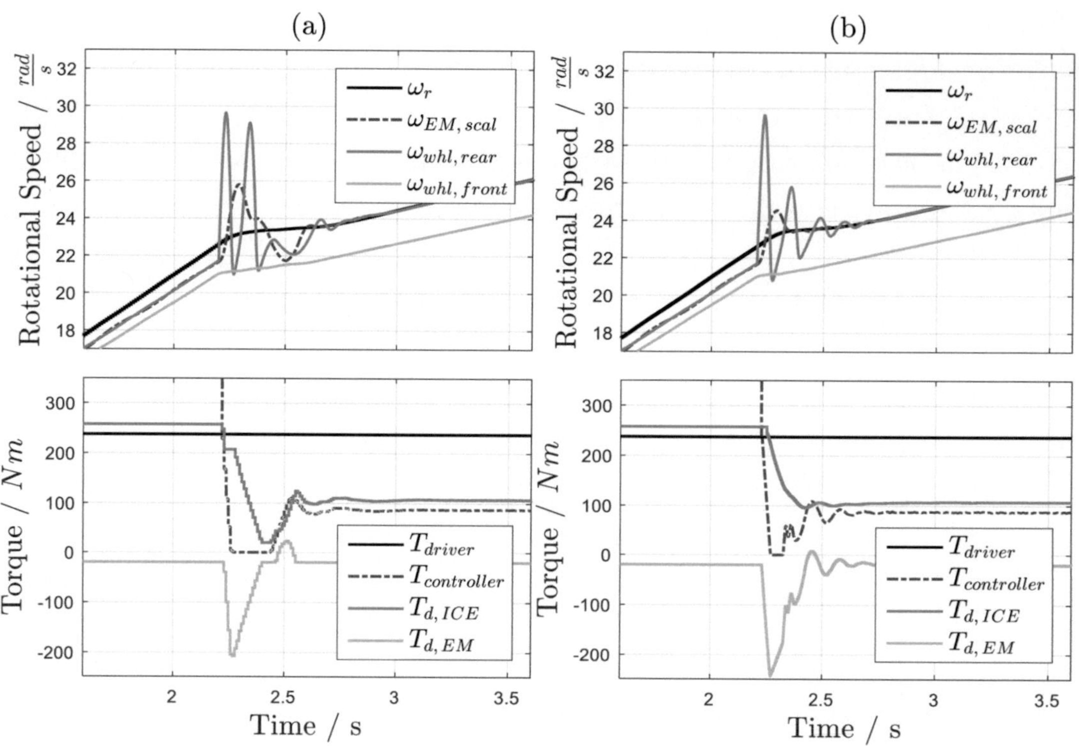

Figure 5.10.: Simulation of Control Allocation for Hybrid Drive of Test Vehicle
Model for Step in Friction Coefficient from High to Low - Design
A (a) and Design B (b)

With Design B on the other hand a clear differentiation between the control frequencies of ICE and EM can be observed. The ICE receives only low dynamic signals filtered by the previously designed low pass. No significant over or undershoots occur as the filter has no oscillatory behavior. The EM receives the dynamic part of the frequency spectrum as defined by the DCA. The steady state value is reached shortly after the controlled variable reaches its steady state value at 2.4 s. Again, a higher oscillatory behavior is observed which can be traced back to the incorporation of a damping term in the controller. Considering the properties of the ECU network and actuators involved the optimization approach yields a highly performant parameterization for the TC problem.

Both implementation designs show desirable functionality and good control performance. Also, the CA designs perform as desired in their given environment. After verifying both full systems in a simulation, the validation in the test vehicle can be carried out.

6 Validation and Evaluation

This chapter deals with benchmarking both proposed TC concepts for HEVs developed in the previous chapters. First, the test procedure and performance measures will be discussed. This is followed by a depiction of the test vehicle, the integration of the TC concepts, as well as the measurement configuration. Then, measurement results of the selected four maneuvers will be depicted for hybrid drive. Based on these, the TC concepts are benchmarked.

6.1. Test Procedure

6.1.1. Driving Maneuvers

To begin with, known publications on TC test procedure will be discussed. Schweers correlates measurements of traction control maneuvers with subjective ratings by test subjects [121]. In further works an objective method to benchmark traction controllers is proposed [120]. Two maneuvers are proposed. The first describes a full throttle acceleration on a low friction surface starting from low and medium vehicle velocities, 5 and 40 km/h respectively. The second maneuver is driving a circle with a constant steering wheel angle with step-wise increases of the gas pedal position. For evaluation performance measures for the individual maneuvers such as yaw angle or longitudinal acceleration are measured at a distinct time after start of control.

More detailed maneuvers are proposed by Pinto [103]. They can be categorized in two categories. In the first category the disturbance on the system is provoked by a step in the requested torque. Here, two slip setpoints are chosen. One close to the maximum and one well above the maximum in order to benchmark the behavior if the target value is uncertain and well above the maximum traction potential. The second category provokes disturbances on the system by a changing friction coefficient. This produces higher control errors and therefore leads to a higher control action posing higher stress on the system.

Most publications focus on longitudinal dynamics since this is the main objective of traction controllers. Additionally, lateral maneuvers with combined control action of the vehicle controller stabilizing the vehicle movement through individual wheel braking and the traction controller are discussed in publications [41]. This also addresses maneuvers with a split friction coefficient provoking combined control action of the traction controller and a brake controller [146]. Since this work focusses solely on the traction controller maneuvers that only address this are chosen. This choice is also motivated by the core of this work focusing on an improved control structure regarding torque distribution and control dynamics. Four distinct maneuvers are proposed for benchmarking the two control system designs. All maneuvers will be performed in second gear as it still provides enough torque for a loss of traction on medium friction surfaces. During these, environment conditions are only known to a limited degree and are subject to stochastic variation.

μ-Step - High to Low

Step in friction coefficient from high to low during high acceleration. This maneuver depicts the most stressful to the controller since much excess torque has to be reduced quickly to prevent the excess wheel slip. Starting at a combined torque of ICE and EM of 450 Nm on dry asphalt $\mu_{asphalt,dry} \approx 1$ the friction coefficient reduces abruptly as the vehicle enters a snow-covered surface $\mu_{snow} \approx 0.2$. This occurs at vehicle speeds of $v_{step} \approx 20$ km/h.

Tip-In

Sudden actuation of the accelerotor pedal to 100 % from constant speed on a low friction coefficient. This maneuver occurs when driving on low friction surfaces such as ice, snow, or wet roads when the driver misjudges the environment conditions by demanding too much torque. The maneuver will start at a constant speed of $v_{initial} \approx 20$ km/h when the driving pedal is pressed to $\alpha_{pedal} = 100$ % abruptly. The experiment is performed on a snow-covered road.

Steps in Reference Value

Steps in the reference slip value. This maneuver shows the response performance of the control system to changes in the setpoint. It is useful when designing the outer loop of the traction controller. The experiment is conducted on a synthetical surface, watered metal planks. These have a friction coefficient of

$\mu_{metal,wet} \approx 0.16$. Starting from a steady state control at $\lambda_r = 0.04$ the desired slip is varied at intervals of 1.5 s beginning at $v_{step} \approx 20$ km/h. Given the critical slip value of approximately $\lambda_{crit} \approx 0.05$ the reference slip values are chosen $\lambda_r = [0.04, 0.01, 0.08, 0.14, 0.02, 0.04]$. A large spectrum range was chosen to demonstrate the wide operating range of the developed concepts.

μ-Step - Low to High

Step in friction coefficient from low to high during active traction control. This maneuver shows the ability of the controller to ramp up its limiting torque to deactivate and therefore, transferring the control back to the drivers request. Starting in steady state control on a snow-covered surface and fully pressed driving pedal the vehicle enters dry asphalt. This occurs at $v_{step} \approx 30$ km/h.

A limitation concerning the chosen maneuvers is their generic nature. In real world scenarios many other maneuvers occur and also overlap. The maneuvers though were chosen as they pose a high level of reproducibility which is needed for a comparison and evaluation. They cover the most common maneuvers and the most stressful to vehicle and controller [23, 103].

Although the chosen maneuvers were selected for their high reproducibility, experimental vehicle tests still are subject to stochastical variations. These especially concern road unevenness and variation in the local friction coefficients [126, 93]. Although performing the tests in constant conditions regarding temperature and humidity stochastical variations occur. Furthermore, the maneuvers are subject to the drivers influence. This concerns timing and actuation precision of the accelerator pedal. Also, slight steering cannot be prevented. To minimize these influences, the test environment was chosen carefully so that these nonpreventable circumstances do not affect the test significantly. Also, all maneuvers were performed several times and representative measurements were chosen for display and benchmarking. Despite these limitations, an experimental study was chosen as many authors do not validate their results in real world applications and focus on theoretical observations.

6.1.2. Performance Measures

Measures for TC and control performance in general are listed in [103] and [87] respectively. Relevant measures are selected, and additional proposed.

Mean Acceleration

The longitudinal acceleration is a key performance measure for TC as maximum acceleration is desired. The measure is defined as

$$a_{mean} = \frac{1}{t_{end} - t_{start}} \int_{t_{start}}^{t_{end}} a_x(t)dt \,, \tag{6.1}$$

where t_{start} defines the absolute time at the start of the maneuver and t_{end} the absolute time at the end. For the following study

$$t_{end} = t_{start} + 1 \text{ s} \,. \tag{6.2}$$

This limitation is motivated by the fact that after one second after start of control action steady state conditions are achieved. After this period, the controlled variable settles and similar accelerations for both designs are the outcome.

Maximum Slip

Defines the maximum longitudinal slip value during a maneuver. It usually occurs directly after the disturbance of the chosen maneuver. It is defined by

$$\lambda_{max} = \max_{t} \lambda_x(t) \,. \tag{6.3}$$

Control Error

The control error is a measure for the tracking performance of the inner control loop. The desired or reference speed will be compared to the scaled actual drivetrain speed. Pinto suggests evaluating the root mean square error (RMSE)

$$RMSE = \sqrt{\frac{1}{t_{end} - t_{start}} \int_{t_{start}}^{t_{end}} (e(t))^2 dt} \,. \tag{6.4}$$

The error is defined by

$$e(t) = \omega_r - \frac{\omega_{mot}}{i_{tot}} \,. \tag{6.5}$$

Slip Error

The slip error is a measure for the tracking performance of the outer control loop. Again, the RMSE will be chosen to evaluate the deviation. The error is defined by the difference in desired slip and actual slip

$$e(t) = \lambda_r - \lambda_x. \tag{6.6}$$

Speed Deviation

The speed deviation is a measure for the drivetrain oscillations. The RSME will be evaluated. The error is defined by the difference in scaled actual actuator speed and wheel speed of the driven axle

$$e(t) = \frac{\omega_{mot}}{i_{tot}} - \omega_{whl}. \tag{6.7}$$

Controller Deactivation Time

This value is a measure for the deactivation behavior of the controller. It is defined as the difference in absolute time between controller deactivation and μ-step

$$\Delta t_{deactivation} = t_{deactivation} - t_{\mu\text{-step}}. \tag{6.8}$$

Mean Rise Time

This value shows the response dynamic to changes in reference value. It is defined as the mean 90% rise time of n-steps in the reference value

$$t_{rise,mean} = \frac{1}{n} \sum_{i=1}^{n} t_{rise,90\%,i}. \tag{6.9}$$

6.1.3. Measurement Setup

Specifications of the test vehicle were described in section 3.2. The prototypical setup of the test vehicle is depicted in figure 6.1. A dSPACE MicroAutoBox (MABX) [30] is connected to the CECU by a generic serial interface for fast data transfer at designated interfaces within the high-level functional structure. The reference value generation, as well as the controller of Design A are located on the MABX. The CECU is equipped with series software. It communicates with a standard vehicle CAN with the EECU. Here though, existing signals were manipulated to communicate additional signals between the ECUs. This concerns the reference value ω_r from CECU to EECU and the maximum ICE torque from EECU to CECU. The EECU uses modified software that contains the functionality of Design B including the fast traction controller and the DCA implemented as linear filters, compare figure 5.6. The ECUs are connected via Application CAN and XETK to an ETAS measurement hardware [34]. This hardware as well as the MABX are connected to a single measurement PC for time synchronous measuring. With this setup all relevant data can be measured and the TC functions on the various ECUs can be parameterized.

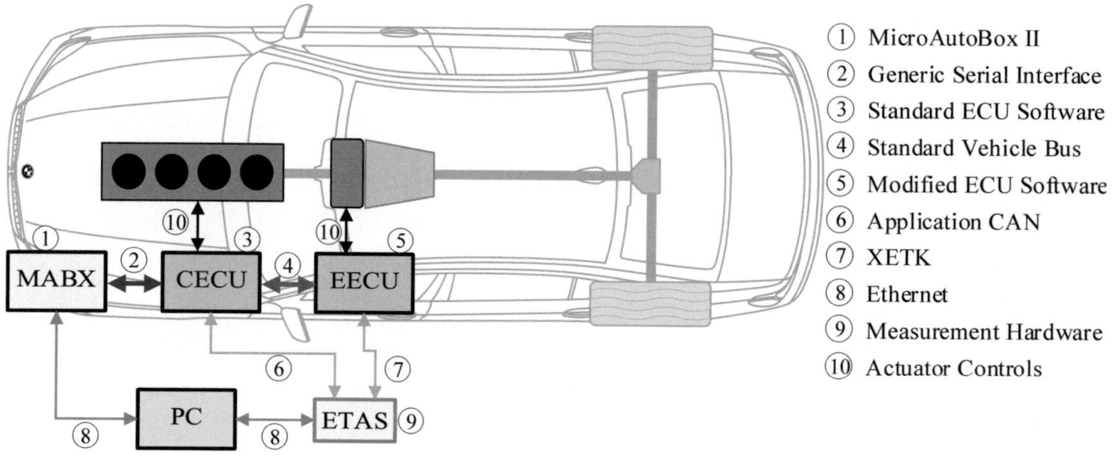

Figure 6.1.: Setup of Test Vehicle

This setup includes a real time rapid control prototyping hardware such as an MABX. This does not reflect the implementation of an industrial vehicle in series production. Also, the manipulation of signals on the bus requires the deactivation of additional functionalities that are not required for TC. Therefore, the experimental study only reflects a final implementation to a limited degree of accuracy. Due to the carefully selected interface choices, the degree of accuracy is high as a seamless integration and fast communication is enabled, mimicking a possible final industrial implementation [111, 129]. Only the reference value generation integrated on the MABX and therefore the CECU does not reflect the final implementation as it was proposed to be located on the DECU. To increase the degree of accuracy of the final implementation a time delay was incorporated for Design A which depicts the delay of bus communication. For Design B this is given by the present bus delay from CECU to EECU.

6.2. Experimental Study and Benchmarking of Control Concepts

Test Bench

Before testing both proposed TC structures in experiments on the road which pose a safety hazard, a verification on a Vehicle in the Loop (ViL) test bench was carried out. The test bench environment of the *Department Automotive Engineering* at the *Technical University of Berlin* was used. With this setup, complete vehicles can be tested in dynamic driving situations by simulating the tire force and exerting it to the wheels by EMs. It is built up of four individually controlled EMs that are connected to the wheel shafts of the vehicle. With these, the ViL test bench can simulate longitudinal and lateral driving maneuvers, and depict the fast dynamics of nonlinear tire road adhesion.

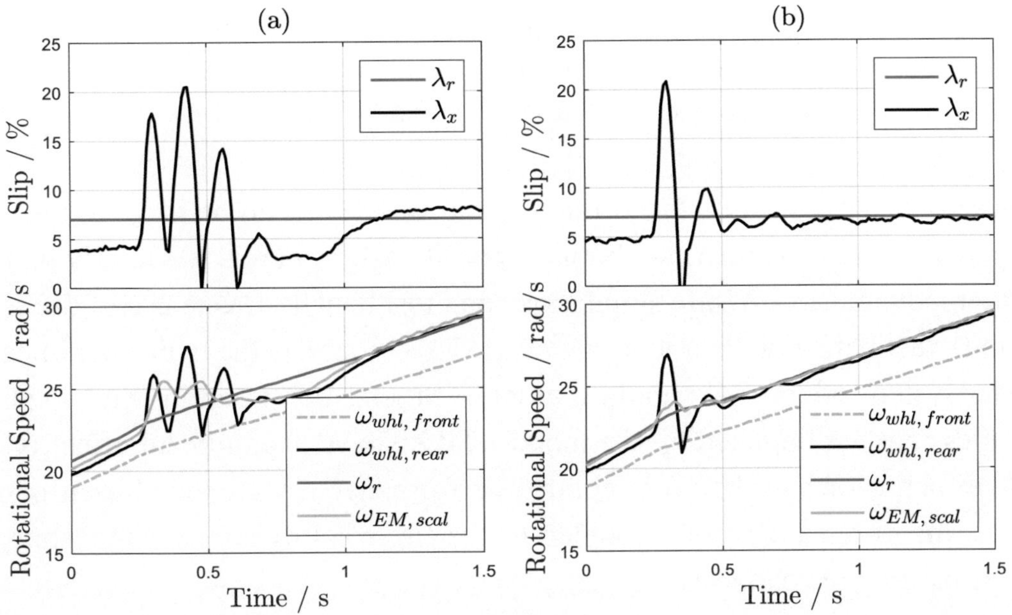

Figure 6.2.: ViL Benchmark: μ-Step - High to Low, Hybrid Drive - Design A (a) and Design B (b) - adapted from [6]

A test bench study for a μ-step from high to low was carried out with the test vehicle. Design A and Design B were compared regarding selected performance measures, see figure 6.2. The top plots depict rotational speeds of interest whereas the bottom plots show reference and desired slip values. For both designs the step in friction coefficient occurs at 0.25 s. The same behavior of the full system compared to the simulation in section 5.4 was observed. Design A shows a high slip and rotational speed overshoot. Distinct oscillations starting with the μ-step occur. Also, an undershoot of the drivetrain speeds and slip can be observed. Steady state conditions settle in at 1 s after the disturbance. Design B on the other hand shows only on distinct overshoot in wheel speed and slip. Steady state conditions settle in at 0.25 s after the disturbance. The results were published in [6]. A quantitative evaluation and comparison can be found in table 6.1. Design B shows improvements in all performance measures except for the max slip value. It remains approximately the same for both designs.

Table 6.1.: Evaluation of Test Bench Measurements for μ-Step - High to Low

Hybrid Drive	Design A	Design B	Delta
Mean Acceleration	1.70 $\frac{m}{s^2}$	1.79 $\frac{m}{s^2}$	+5.1 %
Max. Slip	20.5 %	20.8 %	+1.5 %
RMSE Control	0.78	0.15	−80.5 %
RMSE Slip	8.23	7.48	−9.1 %
RMSE Speed	0.96	0.68	−28.4 %

Based on these results, the validation on road tests can be carried out. The main reason for benchmarking both proposed designs in real world driving tests is tire road adhesion. While simulations and test benches make use of tire models the exact behavior cannot be achieved. This though is the major nonlinearity of the system which TC should regulate. Stochastic variations and road unevenness are not depicted by the model. These have a significant effect on the system behavior and therefore control performance. Also, the chosen performance measures are closely coupled to the tire-road dynamics supporting the benchmarking of road tests.

Road Test

In the following, measurements of all four proposed maneuvers are presented. These were performed with the introduced test vehicle in hybrid drive [1].

Control Design A will be presented in (a) whereas the Design B will be depicted in (b) of the figures. The measurements will be depicted with three charts each. The first shows acceleration and actual slip value. The desired slip value is also marked as a reference. The second chart shows rotational speeds. These

---

[1]Measurement results of electric drive are shown in appendix A.5

include the driven as well as non-driven wheel speeds as a reference. Additionally, the desired wheel speed is displayed which the controller will track. Also, the rotational speeds of the EM or ICE[2] are shown. It is scaled down to wheel level by dividing it by the total gear ratio. This value depicts the controlled variable, see section 4.4.2 and 4.5.2. The last chart shows torques. First, the torque request by the driver is displayed. It is the torque which is allocated to the actuators when slip control is inactive. When the traction controller activates, the system is driven by the desired controller torque. This will be distributed to ICE and EM in hybrid drive by CA. Their desired values are also shown in the plot. During pure electric drive the controller torque will be fully distributed to the EM. Therefore, the desired torque of the ICE is zero. Next to the depiction of the most important signals a quantitative evaluation of the performance measures will be made based on the measurements presented.

The measurements were performed given the assumption that the progression of the μ-λ curve is known. This though in general is not given or only to a limited extent. Especially snow is a driving surface that poses high variations on the tire road adhesion and therefore, also the progression of the μ-λ curve. To minimize the error an identification was performed. From this the critical slip as well as target slip was derived.

6.2.1. μ-Step - High to Low

The measurement results for hybrid drive are shown in figure 6.3. High initial accelerations of 4.5 $\frac{m}{s^2}$ are observable for both concepts. Due to the sudden change in friction coefficient the accelerations drop significantly at 0.5 s to a roughly constant value of about 1.2 $\frac{m}{s^2}$. Design A shows a significant reduction at 1 s to 0.25 $\frac{m}{s^2}$ whereas the acceleration for Design B remains roughly constant. The actual slip has an overshoot in both designs lasting longer for Design A. After the overshoot, a reduction to 0 % occurs followed by another overshoot for Design A. Design B on the other hand regulates the slip more accurately which is shown in table 6.2.

A high overshoot and oscillating transient behavior concerning rotational speeds is observed for Design A whereas Design B shows lower overshoots and a faster tracking of the reference value. Also, lower oscillations of the wheel speeds can be observed for Design B.

---

[2]Depending on the investigated Design and driving mode.

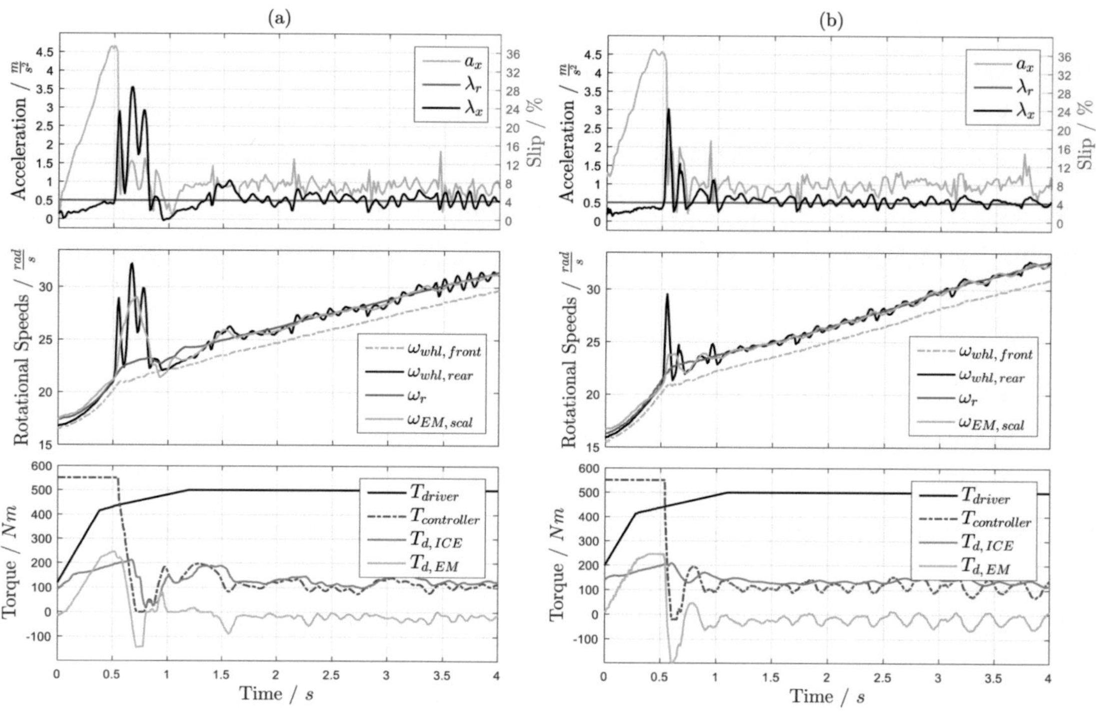

Figure 6.3.: μ-Step - High to Low, Hybrid Drive - Design A (a) and Design B (b)

Table 6.2.: Evaluation of Measurements for μ-Step - High to Low

Hybrid Drive	Design A	Design B	Delta
Mean Acceleration	0.90 $\frac{m}{s^2}$	1.00 $\frac{m}{s^2}$	$+10.7\,\%$
Max. Slip	28.4 %	24.1 %	$-15.1\,\%$
RMSE Control	1.30	0.31	$-76.2\,\%$
RMSE Slip	4.70	2.30	$-51.0\,\%$
RMSE Speed	0.94	0.68	$-27.3\,\%$
Electric Drive			
Mean Acceleration	0.99 $\frac{m}{s^2}$	1.42 $\frac{m}{s^2}$	$+42.2\,\%$
Max. Slip	14.9 %	8.2 %	$-45.0\,\%$
RMSE Control	0.82	0.20	$-76.3\,\%$
RMSE Slip	2.70	0.95	$-64.7\,\%$
RMSE Speed	0.56	0.26	$-53.1\,\%$

The torque reduction by the controller has a steeper gradient in Design B. Additionally, a different behavior in torque distribution is observed. Design A allocates higher frequencies to the ICE which results in an oscillating behavior as opposed to Design B. Steady state values remain similar in both concepts.

6.2.2. Tip-In

The measurement results for hybrid drive are shown in figure 6.4. Quantitative performance measures are depicted in table 6.3. The initial acceleration of 0 $\frac{m}{s^2}$ increases significantly at 0.7 s as the driver requests full torque. The acceleration rises to roughly 2 $\frac{m}{s^2}$ and remains here with small deviations after control begin and settling. A drop to 1.2 $\frac{m}{s^2}$ during settling can be observed for Design A whereas Design B only drops to 1.6$\frac{m}{s^2}$. Also, Design A produces a high slip overshoot to 12.1 % as well as undershoot and significant control errors whereas the Design B regulates the slip more accurately without significant over or undershoot achieving a maximum slip of $\lambda_{max} = 5.9$ %.

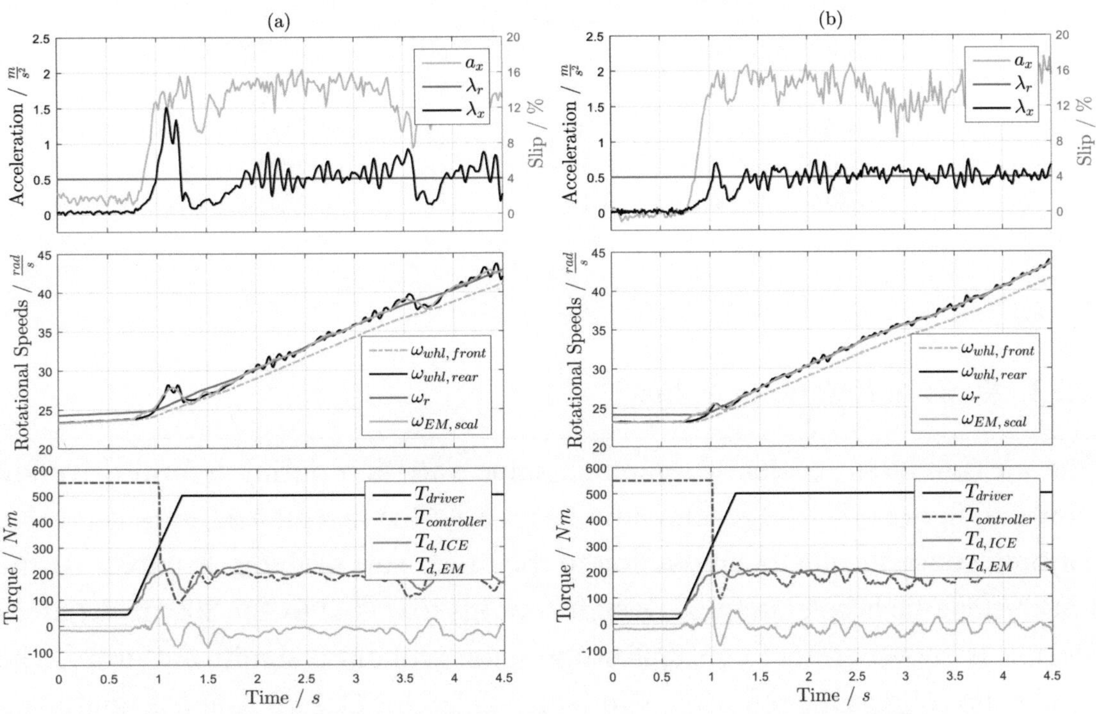

Figure 6.4.: Tip-In, Hybrid Drive - Design A (a) and Design B (b)

Regarding rotational speeds, Design A shows higher control deviations than Design B, see control error in table 6.3. Also, more torsional oscillations are present in Design A. It handles a road unevenness at 3.5 s with higher control deviations than Design B. Here, the unevenness occurs at 2.7 s but no control deviation in rotational speeds is observable.

Design A shows a smoother controller torque progression whereas Design B has a more dynamic control effort. Design A again addresses the ICE with higher dynamics although deviations from the mean value are not as distinct as during the first maneuver. For Design B the steady state torque of the ICE is hardly varied.

Table 6.3.: Evaluation of Measurements for Tip-In

Hybrid Drive	**Design A**	**Design B**	**Delta**
Mean Acceleration	1.64 $\frac{m}{s^2}$	1.83 $\frac{m}{s^2}$	$+12.0\,\%$
Max. Slip	12.1 %	5.9 %	$-51.1\,\%$
RMSE Control	0.61	0.30	$-51.7\,\%$
RMSE Slip	2.25	1.45	$-35.4\,\%$
RMSE Speed	0.36	0.27	$-24.2\,\%$
Electric Drive			
Mean Acceleration	1.47 $\frac{m}{s^2}$	1.80 $\frac{m}{s^2}$	$+22.7\,\%$
Max. Slip	13.9 %	5.4 %	$-61.1\,\%$
RMSE Control	0.59	0.17	$-71.6\,\%$
RMSE Slip	2.45	1.22	$-50.0\,\%$
RMSE Speed	0.19	0.18	$-4.7\,\%$

6.2.3. Steps in Reference Value

The measurement results of reference value variation during hybrid drive are shown in figure 6.5. The acceleration drops after the first step in reference value for both concepts due to a reduction to the linear part of the μ-λ-curve. At the fourth step, the acceleration of Design A drops to $0\,\frac{m}{s^2}$ at $6s$ for 200 ms whereas Design B shows a more constant progression without a significant drop. The actual slip follows the reference slip with a delay for Design A and a significant overshoot. Refer to table 6.4 for a quantitative comparison. Design B can handle changes in the setpoint faster and more accurately. Design A shows high slip oscillations especially for high slip values. With the other design these are not as distinct.

The same oscillations are also present in the speed of the driven axle, compare mean speed deviation in table 6.4. Setpoint tracking with Design B is significantly faster which is resembled by the mean rise time. Both concepts show no steady state error. Again, a significant change in the desired torques of the actuators can be observed. The same behavior as for the μ-step and tip-in occurs.

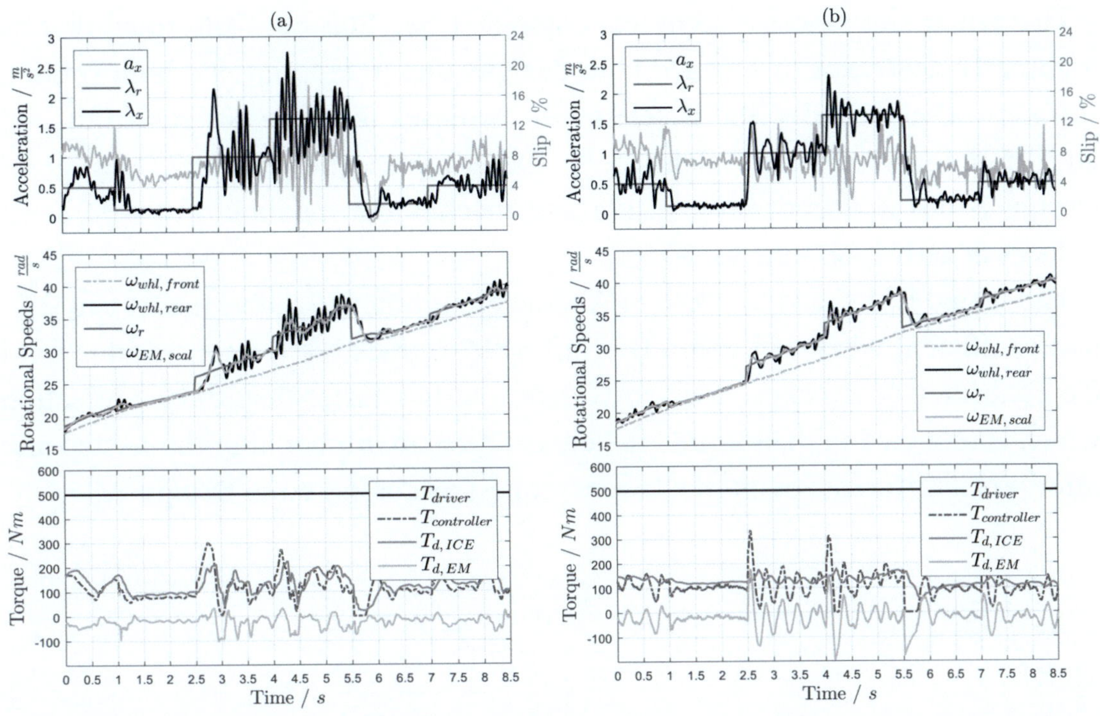

Figure 6.5.: Steps in Reference Value, Hybrid Drive - Design A (a) and Design B (b)

Table 6.4.: Evaluation of Measurements for Steps in Reference Value

Hybrid Drive	**Design A**	**Design B**	**Delta**
Mean Rise Time	0.24 s	0.09 s	$-61.9\,\%$
RMSE Control	0.87	0.58	$-34.0\,\%$
RMSE Slip	2.89	1.85	$-35.9\,\%$
RMSE Speed	0.79	0.45	$-43.3\,\%$
Electric Drive			
Mean Rise Time	0.28 s	0.06 s	$-77.6\,\%$
RMSE Control	1.21	0.46	$-61.9\,\%$
RMSE Slip	3.70	1.63	$-56.0\,\%$
RMSE Speed	1.05	0.49	$-53.4\,\%$

6.2.4. μ-Step - Low to High

Measurement results of a μ-step from low to high during hybrid drive are shown in figure 6.6 to test the controller deactivation behavior. The deactivation criteria for the traction controller was introduced in section 4.1.

The actual slip shows a significant step at $1.5s$. Starting from roughly 4 % it abruptly decreases to 1 % for both designs due to entering the high friction surface. After this step the acceleration increases for both designs from $1\ \frac{m}{s^2}$ until reaching a value of roughly $5\ \frac{m}{s^2}$. Design A reaches it in 0.73 s whereas Design B has a steeper progression and reaches the maximum in 0.34 s. It reduces deactivation time by 52.7 %.

The scaled EM speed for Design A reduces significantly after the step in friction coefficient whereas Design B is still able to track the reference value until the controller deactivates. Design B therefore has faster disturbance rejection which is reflected in the torque gradients. These are higher after the μ-step and offer less interference with the drivers request as compared to Design A.

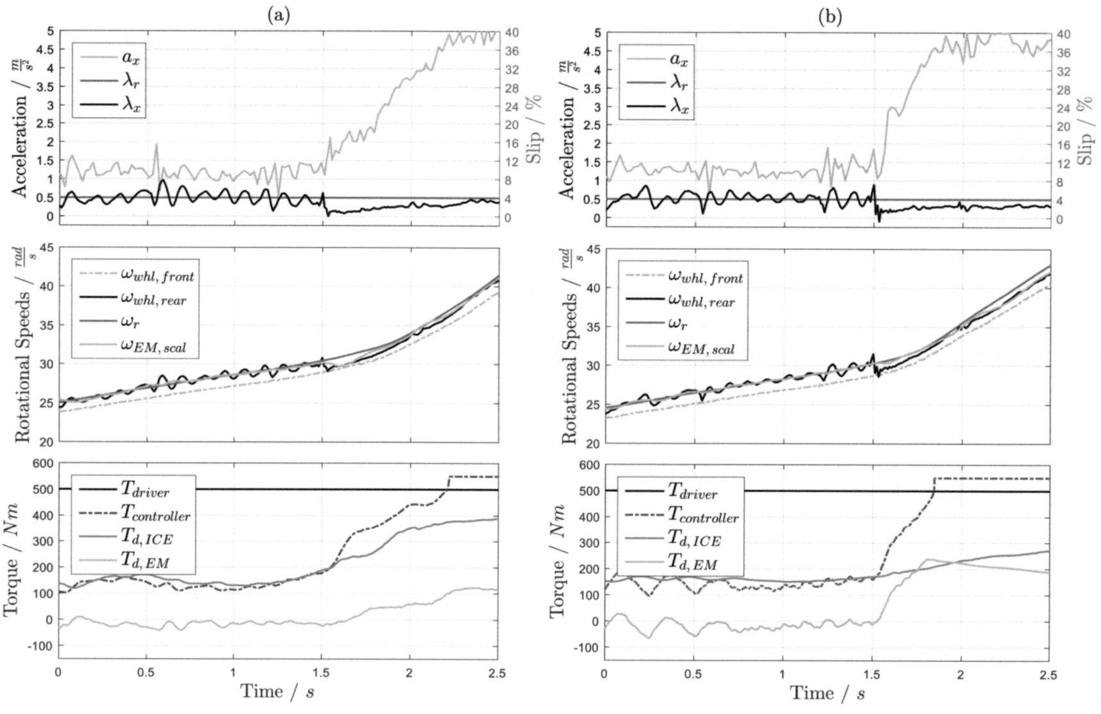

Figure 6.6.: μ-Step - Low to High, Hybrid Drive - Design A (a) and Design B (b)

7 Summary and Outlook

TC is a key safety feature and present in modern industrial vehicles. It prevents the driven wheels from slipping when tractive forces are limited and therefore makes vehicle motion controllable for the driver. Various control structures emerged including speed-based TC which has only briefly been discussed in literature and lacks performant nonlinear approaches. Additionally, new drivetrain technologies gained attention recently for mass produced vehicles driven by strict emission regulations worldwide. A result of these is the increased popularity of electric vehicles and hybrid electric vehicles. The latter use a set of two motors interacting with each other. Applying TC systems to these, again was also only discussed in few works and proposed concepts either lack scalability across technologies and or integrability or allocation technique with good performance. Given the literature review, first the objectives of this work were presented. These include the design of novel nonlinear controllers for speed-based TC and an integration proposal including the use of CA.

To achieve the objectives, first a model of the plant was developed. For this, the nonlinear tire-road adhesion characteristic was included using a common description introduced by Pacejka. Then, the drivetrain oscillatory behavior was analyzed and its description level motivated. Two significant elasticities were identified. These are the half shafts and the two-mass flywheel. Thereafter, the actuators have been described and their dynamic behavior identified for the presented test vehicle. A verification of the complete vehicle model was depicted. It matched the measurements to a high degree of accuracy especially concerning the oscillation behavior.

For the traction controller design, first requirements have been outlined. These were grouped into three categories: development, functionality, and the driver. Based on these, various design specifications for the controllers were derived and the synthesis method of IOL motivated. An example of closed-loop stability for oscillation damping during TC motivated the design of two different controllers, one for slow actuators as active damping leads to an unstable system and one for fast actuators including active damping. The first assumes a rigid drivetrain which requires the implementation of an observer for this virtual system state. The second controller assumes an elastic drivetrain which leads to a damping term when applying IOL. Next to the derivation of two linearizing laws, PID controllers were designed for hybrid and electric drive for both, fast and slow concepts. The results were compared in a simulation showing superior control performance of the fast actuator controller due to its higher open loop bandwidth of factor 5. The controller for slow actuators on the other hand is insensitive to drivetrain oscillations due to the observer enhancing its control performance.

Next, the actuator responses were analyzed when controlled from either powertrain ECU. Controlling the motors from the CECU leads to a similar dynamic response. Controlling from the EECU however, shows a difference in dynamic response by one magnitude order. This led to the proposal of the daisy chain method for CECU and the dynamic control allocation for EECU. This was parameterized with a novel objective method based on an optimization process for a given maneuver. Thereafter, the dynamic properties of each concept including actuators were outlined. Based on these results and the previously designed controllers, two implementation proposals were stated. Additionally, the integration of the additional software to the ECUs and bus was analyzed. The integration was rated to have a minimal impact on the computational intensity and bus load. The implementation concepts were verified in a simulation study. Undesired torque oscillations of the ICE torque for the daisy chain method of Design A were observed. These were not present with DCA in Design B.

A quantitative comparison of both concepts was performed in an experimental study. First, the test procedure including maneuvers and performance measures were motivated and described. This was followed by a depiction of the test vehicle and its measurement as well as prototypical controller implementation setup. It was reasoned to be a close representation of an industrial implementation. Four maneuvers including various steps in reference value as well as two different disturbances from driver and environment were performed. Again, the fast actuator concept outperformed the slow concept. This was shown in the acceleration potential rise of up to 12%, the average rise time reduction of up to 61.9% and the reduction in drivetrain oscillations of up to 43.3%. Also, the parameterized CA for the fast actuator concept showed the desired behavior as the slow actuator was only addressed in the low frequency spectrum.

Many open research topics have been covered with this work. Nevertheless, several questions remain. A promising application of speed-based TC was discussed in [33, 129]. It concerns further control objectives that can be realized with speed-based TC. These are engine drag torque control and hill hold assist. They can be easily realized with the same control structure as reference values for these control tasks are formulated as wheel speeds. In the cited work only drivetrains with one motor have been discussed. What remains open is the influence of a hybrid drivetrain upon these extended control objectives. Special focus also must be set on the control transition between the several tasks.

As mentioned in the limitations, not all real-world driving scenarios were covered in the present work. Further scenarios concern gear shifts from a general viewpoint and the activation and deactivation of the ICE during active TC especially for hybrid drivetrains. The latter must realize a transition of the requested torque in a supervisory control. Additionally, the consequences of an actuator failure can be studied. The presented controllers were designed for fixed gears. Gear shifts though alter the drivetrain behavior. This calls for an extension of the design model and therefore the control law as the gear ratio was assumed to be fixed. Further driving maneuvers concern lateral motion. As TC is relevant for cornering also a comparison of the controllers in these driving maneuvers is relevant for further research. This though requires an adaptation of the reference value generation. It was derived for longitudinal motion only given the slip ratio definition.

Another open field that needs to be addressed in the future is the application of the proposed control concepts to other hybrid topologies that propel one axle with two motors such as power split hybrids. Among parallel hybrids, these are another largely represented group of HEVs. As the powertrain setup offers more DOFs secondary objectives such as a reduction of energy consumption might be a promising research direction.

With this work the key to performant TC was confirmed to be, next to a subtle control design, fast actuator dynamics. This surely can be expanded to vehicle dynamics control. For this, more actuators are present, namely the hydraulic brakes. Also taking other powertrain topologies into account, i.e. vehicles with two driven axles or individually driven wheels, even more DOFs occur. Handling all these actuators with distributed systems as current vehicles are equipped with, only low control dynamics due to communication delays is possible. Enhancing performance is only possible with a structural change: Vehicle dynamics control as well as torque control must be located on a single central ECU. All actuators must be connected to this central ECU with very short time delays < 5 ms. This ensures central and overall control of the vehicle with high performance in stable and unstable driving conditions enabling the optimization of subsequent objectives such as energy recuperation.

Further research can also be done by estimating the friction coefficient before reaching a given road section. This information can be used to limit the driver torque demand and therefore, preventing the vehicle from losing traction in the first place. Promising results on this matter have already been published in [93, 94].

 **Appendix**

A.1. Advanced Input-Output Linearization

This section extends the description in chapter 2.3.4. First, the basic idea of IOL is presented, then, a general approach for a wider range of systems is described. All considerations can also be applied to multiple-input multiple-output systems.

A.1.1. General Concept

Consider the nonlinear system description given in 2.17. If the system is given in canonical form the linearizing law can be determined directly. If this is not the case the system needs to be transformed to its canonical form. For this first the derivative of the system output is calculated. $\dot{y}$ is given by

$$\dot{y} = \frac{\partial h}{\partial x}[f(x) + g(x)u] = L_f h(x) + L_g h(x)\,u$$

where

$$L_f h(x) = \frac{\partial h}{\partial x} f(x)$$

is called the *Lie Derivative* of h with respect to f. This is a convenient notation if the calculation of the derivative is repeated often. It extends to

$$L_f^2 h(x) = L_f L_f h(x) = \frac{\partial(L_f h)}{\partial x} f(x)\,,$$

$$L_f^k h(x) = L_f L_f^{k-1} h(x) = \frac{\partial(L_f^{k-1} h)}{\partial x} f(x)\,.$$

If $L_g h(x) = 0$, then $\dot{y}$ is independent of u. If this is the case a further derivative is calculated. This is done until $L_g L_f^{\rho-1} h(x) \neq 0$ which yields following differential equation

$$y^{(\rho)} = L_f^\rho h(x) + L_g L_f^{\rho-1} h(x)\,u\,.$$

For $\rho = n$, where $n = \dim(\boldsymbol{x})$, the following new state coordinates are defined

$$\boldsymbol{z} = \begin{bmatrix} z_1 \\ z_2 \\ \vdots \\ z_n \end{bmatrix} = \begin{bmatrix} y \\ \dot{y} \\ \vdots \\ y^{n-1} \end{bmatrix} = \begin{bmatrix} c(\boldsymbol{x}) \\ L_f c(\boldsymbol{x}) \\ \vdots \\ L_f^{n-1} c(\boldsymbol{x}) \end{bmatrix} = \boldsymbol{t}(\boldsymbol{x}) \,. \tag{A.1}$$

If $\boldsymbol{t}(\boldsymbol{x})$ and its existing reverse function $\boldsymbol{t}^{-1}$ are continuously differentiable, then

$$\boldsymbol{t}^{-1}(\boldsymbol{t}(\boldsymbol{x})) = \boldsymbol{x} \,, \tag{A.2}$$

forming a diffeomorphism. This is the coordinate transformation looked for. It transforms the system defined in equations 2.17 through differentiation of the transformation equation A.1,

$$\dot{\boldsymbol{z}} = \dot{\boldsymbol{t}}(\boldsymbol{x}) = \frac{\partial \boldsymbol{t}(\boldsymbol{x})}{\partial \boldsymbol{x}} \dot{\boldsymbol{x}} = \begin{bmatrix} L_f c(\boldsymbol{x}) \\ \vdots \\ L_f^{n-1} c(\boldsymbol{x}) \\ L_f^n c(\boldsymbol{x}) L_g L_f^{n-1} c(\boldsymbol{x}) \end{bmatrix} \tag{A.3}$$

explicitly into the nonlinear canonical form

$$\begin{bmatrix} \dot{z}_1 \\ \vdots \\ \dot{z}_{n-1} \\ \dot{z}_n \end{bmatrix} = \begin{bmatrix} z_2 \\ \vdots \\ z_n \\ L_f^n c(\boldsymbol{x}) \end{bmatrix} + \begin{bmatrix} 0 \\ \vdots \\ 0 \\ L_g L_f^{n-1} c(\boldsymbol{x}) \end{bmatrix} u \,, \tag{A.4}$$

$$y = z_1 .$$

Now the control law from this canonical form can be derived. It leads to a linear input-output behavior and is defined by

$$u(\boldsymbol{x}, w) = -r(\boldsymbol{x}) + V w \tag{A.5}$$

where

$$r(\boldsymbol{x}) = \frac{L_f^n c(\boldsymbol{x}) + \boldsymbol{k}^T \boldsymbol{z}}{L_g L_f^{n-1} c(\boldsymbol{x})} \,, \quad \boldsymbol{k}^T = [a_0 \, a_1 \cdots a_{n-1}] \tag{A.6}$$

and

$$v(\boldsymbol{x}) = \frac{V}{L_g L_f^{n-1} c(\boldsymbol{x})} \tag{A.7}$$

as a prefilter. The structure of the resulting system is depicted in figure 2.7. It has linear behavior from the new virtual input w to the output y. This also shows that the diffeomorphism can be interpreted as an input variable transformation being independent from the state vector. Inserting the control law A.5 into A.4 a linear control system is shaped:

$$\begin{bmatrix} \dot{z}_1 \\ \dot{z}_2 \\ \vdots \\ \dot{z}_{n-1} \\ \dot{z}_n \end{bmatrix} = \begin{bmatrix} 0 & 1 & 0 & \cdots & 0 \\ 0 & 0 & 1 & \cdots & 0 \\ \vdots & \vdots & \vdots & \ddots & \vdots \\ 0 & 0 & 0 & \cdots & 1 \\ -a_0 & -a_1 & -a_2 & \cdots & -a_{n-1} \end{bmatrix} \begin{bmatrix} z_1 \\ z_2 \\ \vdots \\ z_{n-1} \\ z_n \end{bmatrix} + \begin{bmatrix} 0 \\ 0 \\ \vdots \\ 0 \\ V \end{bmatrix} w , \tag{A.8}$$

$$y = z_1 .$$

The coefficients a_i of the vector $\boldsymbol{k}^T$ can be chosen freely to place the eigen values of the transformed system to meet desired dynamics. Also, V can be selected freely. With this transformation the nonlinear system input-output behavior was turned to a linear relationship. The differential equation for the system output is given by

$$y^{(n)} + a_{n-1} y^{(n-1)} + \cdots + a_1 \dot{y} + a_0 y = V w . \tag{A.9}$$

From this linear behavior between input w to output y the name *exact input-output linearization* is derived. The schematic is shown in 2.7.

A.1.2. Plants with Internal Dynamics

The initial assumption was that $\rho = n$. Now systems with a lower relative than system degree, $\rho < n$, are discussed. Given a system of the form (2.17) the new state variables are defined as

$$
z = \begin{bmatrix} z_1 \\ z_2 \\ \vdots \\ z_\rho \\ z_{\rho+1} \\ \vdots \\ z_n \end{bmatrix} = t(x) = \begin{bmatrix} c(x) \\ L_f c(x) \\ \vdots \\ L_f^{\rho-1} c(x) \\ t_{\rho+1}(x) \\ \vdots \\ t_n(x) \end{bmatrix} . \tag{A.10}
$$

This shows that the procedure for the first ρ elements is the same as for $\rho = n$. The functions $t_{\rho+1}, \cdots, t_n$ can be chosen freely al long as t form a diffeomorphism. Equations (A.1) and (A.2) must still apply. Transforming the system (2.17) with the diffeomorphism leads to

$$
\dot{z} = \begin{bmatrix} \dot{z}_1 \\ \dot{z}_2 \\ \vdots \\ \dot{z}_{\rho-1} \\ \dot{z}_\rho \\ \dot{z}_{\rho+1} \\ \vdots \\ \dot{z}_n \end{bmatrix} = \begin{bmatrix} L_f c(x) \\ L_f^2 c(x) \\ \vdots \\ L_f^{\rho-1} c(x) \\ L_f^\rho c(x) + L_g L_f^{\rho-1} c(x) u \\ t_{\rho+1}(x) \\ \vdots \\ t_n(x) \end{bmatrix} = \begin{bmatrix} z_2 \\ z_3 \\ \vdots \\ z_\rho \\ L_f^\rho c(x) + L_g L_f^{\rho-1} c(x) u \\ t_{\rho+1}(x) \\ \vdots \\ t_n(x) \end{bmatrix} \tag{A.11}
$$

where

$$
y = z_1
$$

still applies. Here just the first ρ rows are in canonical form as opposed to $\rho = n$ where the entire system is in canonical form. Given the functions $t_{\rho+1}, \cdots, t_n$, t_i is chosen so that

$$
L_g t_i(x) = \frac{\partial t_i(x)}{\partial x} g(x) = 0 . \tag{A.12}
$$

Using the transformation $x = t^{-1}(z)$ the derivatives for $i = \rho + 1, \cdots, n$ are determined. Given (A.12) there is no dependence on u and the derivatives are

$$
\dot{t}_i(x) = L_f t_i(x) = \hat{q}_i(x) = q_i(z) . \tag{A.13}
$$

Equation (A.12) is a partial differential equation. Its solution usually exists but is difficult to determine in many cases. In the following it is assumed that t_i

satisfies equation (A.12) for all $i = \rho + 1, \cdots, n$ and insert $\dot{t}_i$ from (A.13) into (A.11). This results in the *Byrnes Isidori Form*

$$
\begin{bmatrix} \dot{z}_1 \\ \vdots \\ \dot{z}_{\rho-1} \\ \dot{z}_\rho \\ \dot{z}_{\rho+1} \\ \vdots \\ \dot{z}_n \end{bmatrix} = \begin{bmatrix} z_2 \\ \vdots \\ z_\rho \\ \varphi(z,u) \\ q_{\rho+1}(z) \\ \vdots \\ q_n(z) \end{bmatrix}, \quad \left.\begin{array}{c} \\ \\ \end{array}\right\} \text{external dynamics} \\ \left.\begin{array}{c} \\ \\ \end{array}\right\} \text{internal dynamics}
\tag{A.14}
$$

$$
y = z_1,
$$

with the system description

$$
\varphi(z,u) = L_f^\rho c(t^{-1}(z)) + L_g L_f^{\rho-1} c(t^{-1}(z)) u .
\tag{A.15}
$$

The system dynamic can be split up into an external and internal part. The external part which describs the changes of the states $z_1, \cdots, z_\rho$ can be linearized by a transformation of u into the new input w as in the case $\rho = n$. The transformation is the same as in the full rank case and is

$$
u = -\frac{L_f^\rho c(x) + k^T z}{L_g L_f^{\rho-1} c(x)} + \frac{V}{L_g L_f^{\rho-1} c(x)} w
\tag{A.16}
$$

with

$$
k^T = [a_0 \cdots a_{\rho-1} 0 \cdots 0]
\tag{A.17}
$$

Inserting equation (A.16) into (A.14) into the dynamic equations of the transformed system yields

$$
\begin{bmatrix} \dot{z}_1 \\ \dot{z}_2 \\ \vdots \\ \dot{z}_{\rho-1} \\ \dot{z}_\rho \end{bmatrix} = \begin{bmatrix} 0 & 1 & 0 & \cdots & 0 \\ 0 & 0 & 1 & \cdots & 0 \\ \vdots & \vdots & \vdots & \ddots & \vdots \\ 0 & 0 & 0 & \cdots & 1 \\ -a_0 & -a_1 & -a_2 & \cdots & -a_{\rho-1} \end{bmatrix} \begin{bmatrix} z_1 \\ z_2 \\ \vdots \\ z_{\rho-1} \\ z_\rho \end{bmatrix} + \begin{bmatrix} 0 \\ 0 \\ \vdots \\ 0 \\ V \end{bmatrix} w \left. \right\} \quad \text{external} \atop \text{dynamics} \tag{A.18}
$$

$$
\begin{bmatrix} \dot{z}_{\rho+1} \\ \vdots \\ \dot{z}_n \end{bmatrix} = \begin{bmatrix} q_{\rho+1}(\boldsymbol{z}) \\ \vdots \\ q_n(\boldsymbol{z}) \end{bmatrix} \left. \right\} \text{internal dynamics} \tag{A.19}
$$

$$
y = z_1 . \tag{A.20}
$$

Due to the transformation of the input variable the external dynamic becomes independent from the state variables $z_{\rho+1} \cdots z_n$ of the internal dynamic. The state variables of the internal dynamic have no influence upon the system output y. The internal dynamic therefore loses its observability due to the linearization. This is of **negligible significance for applications** [4]. The internal dynamic can be depicted as an independent system with the input variables $z_1 \cdots z_\rho$ and no output. The external dynamic is always observable if

$$
L_g L_f^{\rho-1} c(t^{-1}(\boldsymbol{z})) \neq 0 \tag{A.21}
$$

holds. Again the transformation equation (A.16) is interpreted as a control law with a state space controller and a prefilter with the reference value w. The controlled external dynamic has the same structure as for the case $\rho = n$. The output variable is given by the differential equation

$$
y^{(\rho)} + a_{\rho-1} y^{(\rho-1)} + \cdots + a_1 \dot{y} + a_0 y = V w , \tag{A.22}
$$

but with the rank $\rho < n$. The rank therefore also defines the order of the linear differential equation of the resulting to be controlled external dynamic.

A.2. System Order Reduction

A.2.1. Dominance of Litz

In order to evaluate which eigenvalues of a dynamic system are relevant for the overall dynamic measures of dominance exist. One common dominance measure was introduced by Litz [81, 87]. For this consider a dynamic system of the form

$$\dot{x} = Ax + Bu$$
$$y = Cx\,.$$
(A.23)

Now this system is transformed to by

$$\tilde{x} = V^{-1}x$$
(A.24)

to its canonical form

$$\dot{\tilde{x}} = \tilde{A}\tilde{x} + \tilde{B}u$$
$$y = \tilde{C}\tilde{x}\,.$$
(A.25)

with $A = \mathrm{diag}\lambda_i$. To evaluate each dynamic in the system Litz proposed the following method

$$d_i = \left|\frac{\tilde{b}_i \tilde{c}_i}{\lambda_i}\right|$$
(A.26)

with d_i being the dominance for each eigenvalue λ_i.

A.2.2. Method of Laschet

If the complexity of a system needs to be reduced for simulation or control design a system order reduction is applied. For torsional oscillators Laschet designed a method for this [77]. First the frequency spectrum for the reduced system needs to be defined. Then elements of the system (inertias and stiffnesses and dampers) are eliminated that are only relevant in a higher frequency range and therefore can be neglected. This begins with the stiffness which is responsible for the highest eigen frequency c_i. It lies between two inertias J_n and J_{n+1}. Now J_n will be reduced to the adjacent inertias J_{n-1} and J_{n+1}

$$J'_{n-1} = J_{n-1} + \frac{c_{i-1}}{c_{i-1} + c_i} J_n\,,$$
(A.27)

$$J'_{n+1} = J_{n+1} + \frac{c_i}{c_{i-1} + c_i} J_n\,.$$
(A.28)

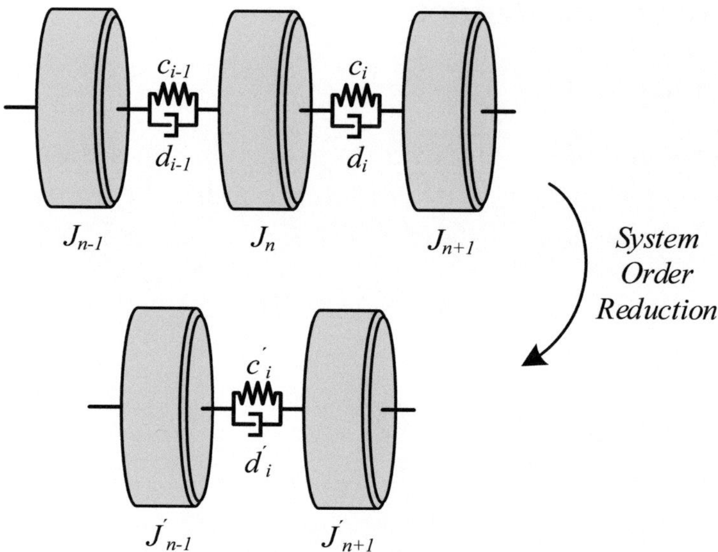

Figure A.1.: System Order Reduction for a Torsional Oscillator using the Method by Laschet

Now the system consists of two inertias and two stiffnesses connected in series. In the final step the two stiffnesses will be reduced to one.

$$c_i' = \frac{c_{i-1}c_i}{c_{i-1}+c_i} \, . \tag{A.29}$$

This equation also holds for the damping coefficients

$$d_i' = \frac{d_{i-1}d_i}{d_{i-1}+d_i} \, . \tag{A.30}$$

The method of system order reduction is depicted in figure A.1. Three inertias connected by two stiffnesses and dampers are reduced to a system of two inertias connected by one stiffness and one damper.

A.3. Model Parameters

Table A.1 shows the parameters used in the simulation model.

Table A.1.: Driving Dynamics Parameters

Parameter	Unit	Value
Air Density	$\frac{kg}{m^3}$	1.27
Drag Coefficient	–	0.25
Friction Coefficient Metal, Wet	–	0.16
Friction Coefficient Snow	–	0.2
Friction Coefficient Asphalt, Dry	–	1
Front Area of Vehicle	m^2	2.42
Gravitational Acceleration	$\frac{m}{s^2}$	9.81
Pacejca Parameter B	–	11
Pacejca Parameter C	–	1.67
Pacejca Parameter E	–	-3
Rolling Friction Coefficient	–	0.01
Tire Radius	m	0.3537
Vehicle Mass	kg	2048.3

Table A.2.: Drivetrain Parameters

Parameter	Unit	Value
Damping Coefficient Flywheel	$\frac{Nms}{rad}$	2
Damping Coefficient Flywheel reduced	$\frac{Nms}{rad}$	1.43
Damping Coefficient Cardan Shaft	$\frac{Nms}{rad}$	5
Damping Coefficient Haf Shaft	$\frac{Nms}{rad}$	11
Damping Coefficient Haf Shaft reduced	$\frac{Nms}{rad}$	9.02
Damping Coefficient Tire Belt	$\frac{Nms}{rad}$	50
Gear Ratio Differential	–	3.077
Gear Ratio Gearbox of Gears [1..8]	–	$[4.71, 3.14, 2.11, 1.67, ...$ $1.29, 1, 0.84, 0.67]$
Rotational Inertia ICE	kgm^2	0.112
Rotational Inertia ICE reduced	kgm^2	0.130
Rotational Inertia EM, incl. Gearbox	kgm^2	0.2069
Rotational Inertia EM reduced	kgm^2	0.193
Rotational Inertia Differential	kgm^2	0.0674
Rotational Inertia Rim	kgm^2	2.1

Parameter	Unit	Value
Rotational Inertia Wheel reduced	kgm^2	3.225
Rotational Inertia Tire Belt	kgm^2	1.36
Rotational Inertia Motor Hybrid reduced	kgm^2	0.341
Rotational Inertia Motor Electric reduced	kgm^2	0.193
Torsional Stiffness Flywheel	$\frac{\text{Nm}}{\text{rad}}$	990
Torsional Stiffness Flywheel reduced	$\frac{\text{Nm}}{\text{rad}}$	1036
Torsional Stiffness Cardan Shaft	$\frac{\text{Nm}}{\text{rad}}$	10000
Torsional Stiffness Half Shaft	$\frac{\text{Nm}}{\text{rad}}$	11328
Torsional Stiffness Half Shaft reduced	$\frac{\text{Nm}}{\text{rad}}$	9558
Torsional Stiffness Tire Belt	$\frac{\text{Nm}}{\text{rad}}$	90000

Table A.5 shows the identified parameters of the actuators. Parameters which are not shown were defined in section 3.2 or 3.3.4 and are subject to a physical relationship.

Table A.3.: Actuator Parameters

Parameter	Unit	Value
Maximum Aspirated Torque ICE	Nm	220
Time Delay Ignition Path	s	$\frac{n_{ign}}{n_{cyl}}\frac{2\pi\frac{rad}{s}}{\omega_{ICE}}$
Time Constant Ignition	s	$\frac{\alpha_{pressure}}{360°}\frac{2\pi\frac{rad}{s}}{\omega_{ICE}}$
Angle Pressure Build Up	°	60
Time Delay Air Path Naturally Aspirated	s	$0.026 + \frac{2.358\frac{rad}{s}}{\omega_{ICE}}$
Time Constant Naturally Aspirated	s	$\frac{3.998\frac{rad}{s}}{\omega_{ICE}}$
Time Constant Turbo Positive	s	$-712 + 8.22\frac{s}{rad}\omega_{ICE}$
Time Constant Turbo Negative	s	$-3534 - 4.13\frac{s}{rad}\omega_{ICE}$
Rate Limitation of EM Torque	$\frac{\text{Nm}}{\text{s}}$	6000
Time Delay EM	s	0.004
Time Constant EM	s	0.0015
Time Constant Red. Design A Hyb. $1500\frac{1}{\text{min}}$	s	0.036
Time Constant Red. Design A Hyb. $5000\frac{1}{\text{min}}$	s	0.021
Time Constant Red. Design A Electric	s	0.0355
Time Constant Red. Design B Hyb.	s	0.0055
Time Constant Red. Design B Electric	s	0.0055

Table A.4.: Control Network Parameters

Parameter	Unit	Value
Sample Time CECU	s	0.01
Sample Time EECU	s	0.001
Time Delay Bus Communication CECU	s	0.03
Time Delay Bus Communication EECU	s	0.02

A.4. Control System Parameters

Table A.5.: Control System Parameters

Parameter	Unit	Value
Proportional Gain, Slow Actuator Hybrid	–	39
Integral Gain, Slow Actuator Hybrid	–	200
Derivative Gain, Slow Actuator Hybrid	–	0.46
Proportional Gain, Slow Actuator Electric	–	16
Integral Gain, Slow Actuator Electric	–	60
Derivative Gain, Slow Actuator Electric	–	0.17
Proportional Gain, Fast Actuator	–	82.8
Integral Gain, Fast Actuator	–	700
Derivative Gain, Fast Actuator	–	0.143
Observer Gain 1 Hybrid	–	$\begin{bmatrix} 120.9 \\ 14.66 \end{bmatrix}$
Observer Gain 2 Hybrid	–	1
Observer Gain 3 Hybrid	–	$\begin{bmatrix} 1200 \\ 53 \end{bmatrix}$
Factor Eigenfrequency Observer Hybrid	–	0.7
Damping Observer Hybrid	–	1
Observer Gain 1 Electric	–	$\begin{bmatrix} 33.61 \\ 40.58 \end{bmatrix}$
Observer Gain 2 Electric	–	1
Observer Gain 3 Electric	–	$\begin{bmatrix} 2000 \\ 80 \end{bmatrix}$
Factor Eigenfrequency Observer Electric	–	1.2
Damping Observer Electric	–	1
Weighting Matrix 1 DCA	–	$\begin{bmatrix} 0.9642 & 0 \\ 0 & 1.0102 \end{bmatrix}$

Parameter	Unit	Value
Weighting Matrix 2 DCA	–	$\begin{bmatrix} 21.0863 & 0 \\ 0 & 0.9923 \end{bmatrix}$
Weighting Matrix v DCA	–	1
Weighting Parameter 1 Quality Function	–	6.4516
Weighting Parameter 2 Quality Function	–	0.6576
Weighting Parameter 3 Quality Function	–	1

A.5. Validation of Electric Drive

μ-Step - High to Low

Figure A.2 shows measurement results for a μ-step from high to low during pure electric drive. Since the EM cannot produce as much torque as the combined system of ICE and EM the maximum torque is reduced and therefore the initial acceleration as well. Starting at 2.7 $\frac{m}{s^2}$ again the friction coefficient is reduced at 0.5 s. Design B shows a rather constant acceleration of about 1.2 $\frac{m}{s^2}$ after the step whereas Design A has more oscillations. Again, the reason for this can be seen

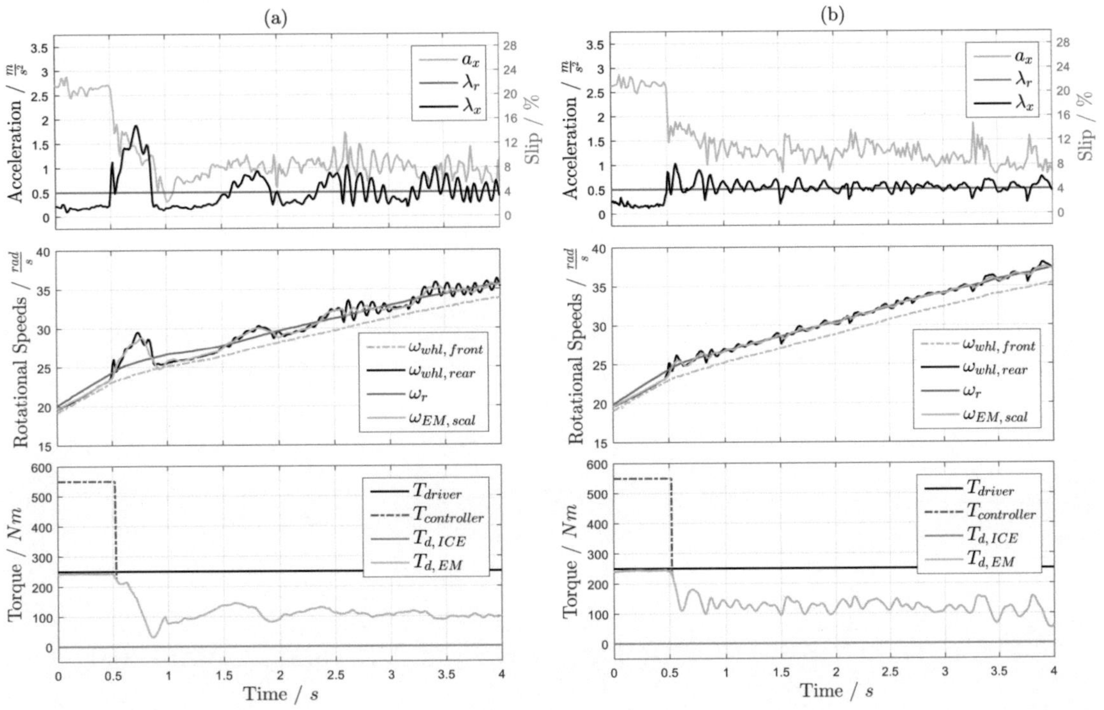

Figure A.2.: μ-Step - High to Low, Electric Drive - Design A (a) and Design B (b)

in the second plot. Low frequency oscillations around the desired values with high overshoot occur in Design A but also high frequency oscillations starting at 2.5 s. The proposed concept is able to track the reference values more accurately showing fewer oscillations.

The last plot shows the desired torque of the controller which is equal to the desired torque of the EM. Design A shows a low torque gradient after the drop in friction coefficient whereas Design B allows a higher gradient and higher frequency torque modulations due to its higher open-loop bandwidth. Further quantitative results are depicted in table 6.2.

Tip-In

Measurement results of maneuver two during electric drive are depicted in figure A.3. The acceleration for Design A shows a significant drop after start of regulation. This can be explained by a high slip undershoot starting at 1.4 s and lasting to 2.9 s. As opposed to this Design B does not produce an overshoot and the undershoot is lower and shorter. This lets the vehicle accelerate at a constant value of about 1.9 $\frac{m}{s^2}$. These observations are backed by the second

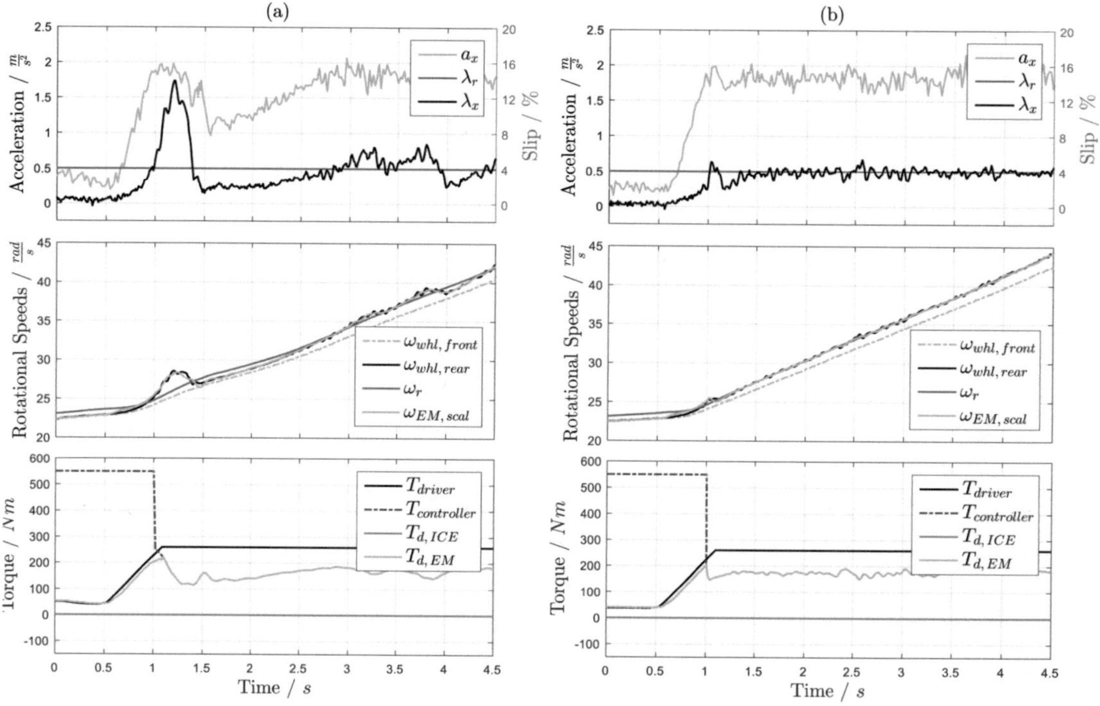

Figure A.3.: Tip-In, Electric Drive - Design A (a) and Design B (b)

plot. Design A produces a high overshoot and a high and long undershoot. The following tracking also shows higher deviations than Design B. The reason for this is shown in the last plot. Design A has a much smoother controller torque porgression than Design B. Further quantitative results are depicted in table 6.3.

Step in Reference Value

The results of setpoint variation during electric drive are depicted in figure A.4. The top two plots are closely related to the ones in hybrid drive. Nevertheless, higher oscillations are abserved for Design A compared to hybrid drive. Additionally, higher overshoots are present. Again a drop of the acceleration to $0 \frac{m}{s^2}$ can be seen at 6 s for Design A which is not present with Design B. The tracking of the setpoint in Design B is faster than in hybrid drive and outperforms of Design A. Quantitative results are depicted in table 6.4.

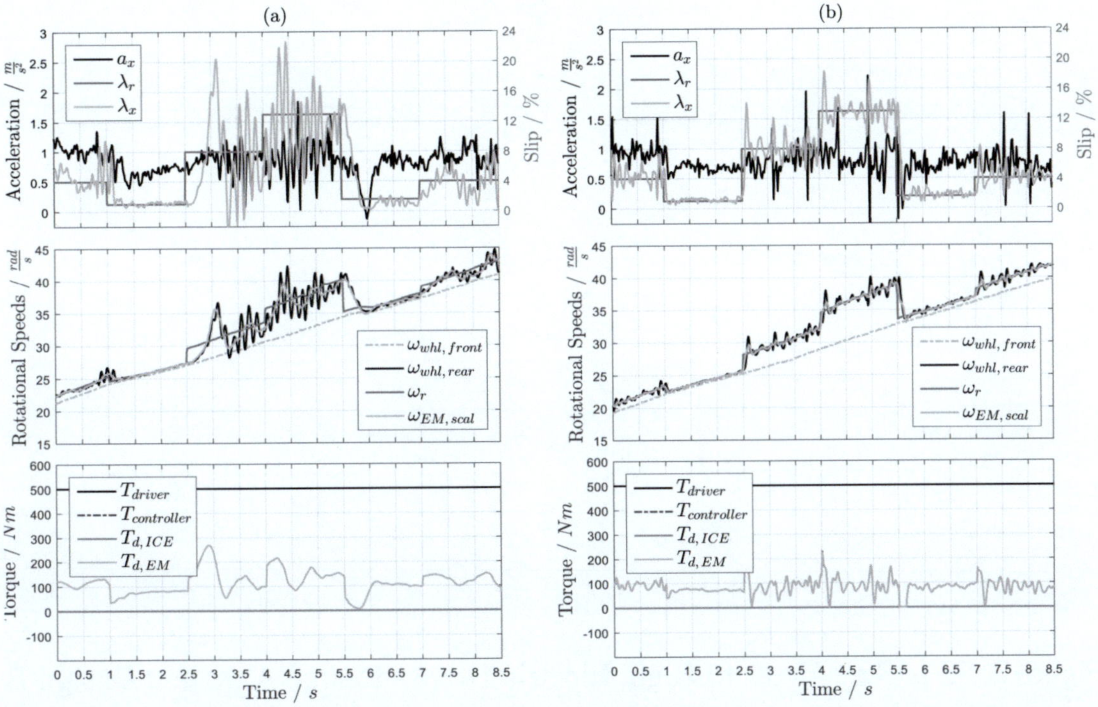

Figure A.4.: Steps in Reference Value, Electric Drive - Design A (a) and Design B (b)

μ-Step - Low to High

The acceleration slope for Design B after the μ-step at 1.5 s is steeper. Therfore, the deactivation time is reduced from roughly 0.52 s for Design A to 0.07 s for Design B. This is a reduction of 87.0 %. The controlled variable for Design B remains at the desired value for the initial 0.2 s after the step whereas Design A is not able to maintain the desired value after the step.

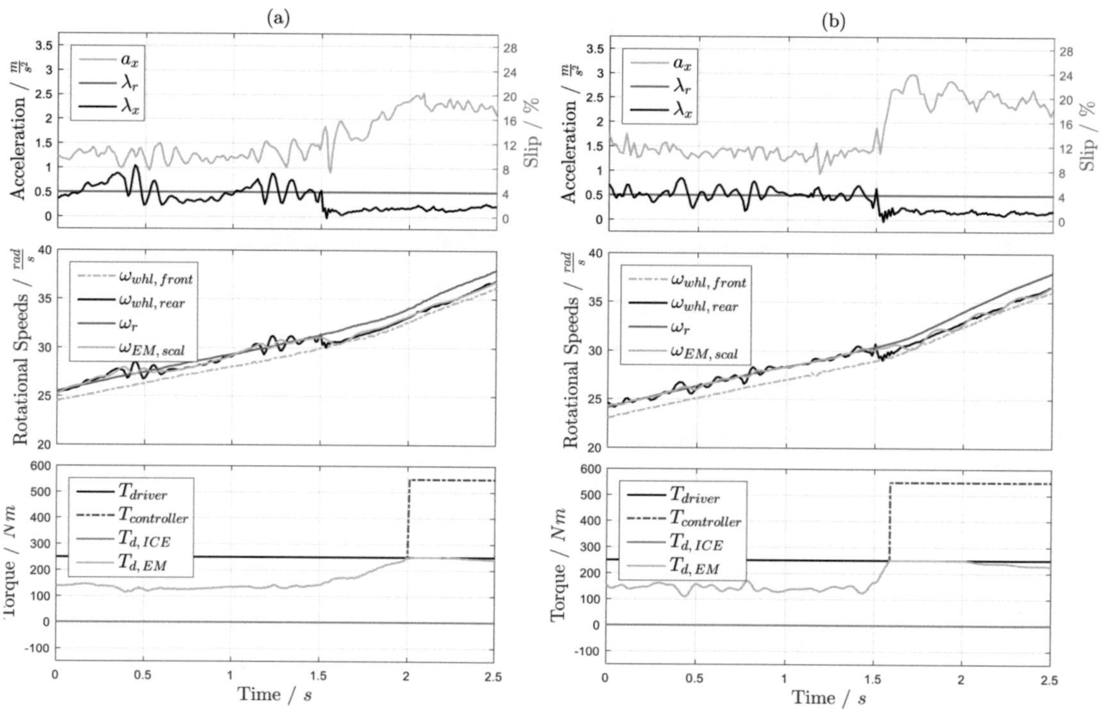

Figure A.5.: μ-Step - Low to High, Electric Drive - Design A (a) and Design B (b)

List of Figures

List of Tables

Bibliography

[1]   *Bosch automotive electrics and automotive electronics: Systems and components, networking and hybrid drive.* 5. ed. Wiesbaden : Springer Vieweg, 2014 (Bosch Professional Automotive Information). – ISBN 978–3–658–01783–5

[2]   ABDELHAMEED, M. ; ABDELAZIZ, M. ; ELHARDY, N. ; HUSSEIN, A. : Development of Integrated Brakes and Engine Traction Control System. In: *International Workshop on Research and Education in Mechatronics* 15th (2014)

[3]   ADAMSKI, D. : *Simulation in der Fahrwerktechnik: Einführung in die Erstellung von Komponenten- und Gesamtfahrzeugmodellen.* Wiesbaden : Springer Vieweg, 2014 (ATZ / MTZ-Fachbuch). – ISBN 978–3–658–06535–5

[4]   ADAMY, J. : *Nichtlineare Systeme und Regelungen.* 2., bearb. und erw. Aufl. Springer Vieweg. http://dx.doi.org/10.1007/978-3-642-45013-6. http://dx.doi.org/10.1007/978-3-642-45013-6. – ISBN 978–3–642–45012–9

[5]   AKIBA, T. ; SHIRATO, R. ; FUJITA, T. ; TAMURA, J. : A Study of Novel Traction Control Method for Electric Motor Driven Vehicle. In: *IEEE Power Conversion Conference* (2007)

[6]   AL-SAIDI, O. ; MÜLLER, S. ; ZECH, A. ; SCHYR, C. :   Bewertung und Analyse von Fahrzeugregelsystemen im fahrdynamischen Grenzbereich mit einem Vehicle-in-the-Loop Prüfstand. In: *VDI-Kongress SIMVEC - Simulation und Erprobung in der Fahrzeugentwicklung* 19th (2018)

[7]   ANTONOV, S. ; FEHN, A. ; KUGI, A. :   Unscented Kalman Filter for Vehicle State Estimation. In: *Vehicle System Dynamics* 49 (2011), Nr. 9, pages 1497–1520

[8]   ANWAR, S. :   An Anti-Lock Braking Control System for a Hybrid Electromagnetic/Electrohydraulic Brake-By-Wire System. In: *Proceedings of the American Control Conference* (2004), pages 2699–2704

[9]   ÅSTRÖM, K. J. ; MURRAY, R. M.: *Feedback Systems: An Introduction for Scientists and Engineers.* Princeton Univ. Press http://www.loc.gov/catdir/enhancements/fy0806/2007061033-b.html. – ISBN 978–0–691–13576–2

[10]   BAKKER, E. ; PACEJKA, H. B. ; LIDNER, L. : A New Tire Model with an Application in Vehicle Dynamics Studies. In: *SAE Transactions - Journal of Passenger Cars* 98 (1989), Nr. 6, pages 101–113

[11]   BALAZS, A. : *Optimierte Auslegung von ottomotorischen Hybridantriebssträngen unter realen Fahrbedingungen.* Aachen, RWTH Aachen, Dissertation, 2015

[12]   BASSHUYSEN, R. ; SCHAEFER, F. : *Handbuch Verbrennungsmotor: Grundlagen, Komponenten, Systeme, Perspektiven.* 7th. Wiesbaden : Springer Vieweg, 2015

[13]   BAUER, H. : *Kraftfahrtechnisches Handbuch.* 22nd. Berlin, Heidelberg : Springer-Verlag, 1995

[14] BEALE, S. ; SHAFAI, B. : Robust Control System Design with Proportional Integral Observer. In: *International Journal of Control* 50 (2007), Nr. 1, pages 97–111

[15] BORDINGTON, K. A.: *Constrained Control Allocation for Systems with Redundant Control Effectors*. Virginia, Dissertation, 1996

[16] BOTTIGLIONE, F. ; SORNIOTTI, A. ; SHEAD, L. : The Effect of Halfshaft Torsion Dynamice on the Performance of a Traction Control System for Electric Vehicles. In: *Journal of Automobile Engineering* 226 (2012), Nr. 9, pages 1145–1159

[17] BROOKS, R. J. ; TOBIAS, A. M.: Choosing the Best Model: Level of Detail, Complexity, and Model Performance. In: *Mathematical and Computer Modelling* 24 (1996), Nr. 4, pages 1–14

[18] BUFFINGTON, J. M. ; ENNS, D. F.: Lyapunov Stability Analysis of Daisy Chain COntrol ALlocation. In: *Journal of Guidance, Control and Dynamics* 19 (1996), Nr. 6, pages 1226–1230

[19] BUSAWON, K. ; KABORE, P. : Disturbance Attenuation Using Proportional Integral Observers. In: *International Journal of Control* 74 (2001), Nr. 6, pages 618–627

[20] CASTRO, R. de ; ARAUJO, R. E. ; FREITAS, D. : Real-Time Estimation of Tyre-Road Friction Peak with Optimal Linear Parameterisation. In: *IET Control Theory & Applications* 6 (2012), Nr. 14, pages 2257–2268

[21] CASTRO, R. de ; ARAUJO, R. E. ; FREITAS, D. : Wheel Slip Control of EVs Based on Sliding Mode Technique with Conditional Integrators. In: *IEEE Transactions on Industrial Electronics* 60 (2013), Nr. 8, pages 3256–3271

[22] CASTRO, R. de ; ARAUJO, R. E. ; TANELLI, M. ; SAVARESI, S. M. ; FREITAS, D. : Torque Blending and Wheel Slip Control in EVs with In-Wheel Motors. In: *Vehicle System Dynamics* 50 (2012), Nr. 1, pages 71–94

[23] CHAPIUS, C. ; BIDEAUS, E. ; BRUN, X. ; MINOIU-ENACHE, N. : Comparison of Feedback Linearization and Flatness Control for Anti-Slip Regulation (ASR) of a Hybrid Vehicle: From Theory to Experimental Results. In: *European Control Conference* (2013), pages 446–451

[24] CHENGLIN, L. ; SHOUBO, L. ; SHANGLOU, C. ; LIFANG, W. : Modelling and Simulation of Traction Control of Hybrid Electric Vehicle. In: *Proceedings of IEEE International COnference on Mechatronics and Automation* (2009), pages 3252–3256

[25] CORDINER, S. ; GALEANI, S. ; MECOCCI, F. ; MULONE, V. ; PERANTONI, G. ; ZACCARIAN, L. : Dynamic Input Allocation of Torque References for a Parallel HEV. In: *IEEE Conference on Decision and Control (CDC)* 49th (2010)

[26] CORDINER, S. ; GALEANI, S. ; MECOCCI, F. ; MULONE, V. ; ZACCARIAN, L. : Torque Setpoint Tracking for Parallel Hybrid Electric Vehicles Using Dynamic Input Allocation. In: *IEEE Transactions on Control Systems Technology* 22 (2014), Nr. 5, pages 2007–2015

[27] DANKERT, J. ; DANKERT, H. : *Technische Mechanik: Statik, Festigkeitslehre, Kinematik/Kinetik ; mit 128 Übungsaufgaben, zahlreichen Beispielen und weiteren Aufgaben im Internet*. 5., überarb. und erw. Aufl. Wiesbaden : Vieweg + Teubner, 2009 (Studium). – ISBN 9783835101777

[28] DEUR, J. ; PAVKOVIC, D. ; BURGIO, G. ; HROVAT, D. : A Model-Based Traction Control Strategy Non-Reliant on Wheel Slip Information. In: *Vehicle System Dynamics* 49 (2011), Nr. 8, pages 1245–1265

[29] DOYLE, J. C. ; FRANCIS, B. A. ; TANNENBAUM, A. R.: *Feedback Control Theory.* Dover Publications (Dover Books on Electrical Engineering). http://gbv.eblib. com/patron/FullRecord.aspx?p=1897347. – ISBN 978–0486469331

[30] DSPACE: *Manufacturer Information on MicroAutoBox.* https://www.dspace. com/de/gmb/home/products/hw/micautob/microautobox2.cfm. Version: 1. Access Date: 06.01.2020

[31] DURHAM, W. C.: Constrained Control Allocation. In: *Journal of Guidance, Control and Dynamics* 16 (1993), Nr. 4, pages 717–725

[32] DURHAM, W. C. ; BORDINGTON, K. A.: Multiple Control Effector Rate Limiting. In: *Journal of Guidance, Control and Dynamics* 19 (1996), Nr. 1, pages 30–37

[33] EBERL, T. ; ESPENHAIN, M. ; BAIER, A. ; MÜLLER, T. : *Drive Dynamic Regulating System in a Motor Vehicle and Electronic Drive Dynamic Control Unit for a Drive Dynamic Regulating System.* 2017

[34] ETAS: *Manufacturer Information on XETK.* https://www.etas.com/de/ portfolio/xetk.php. Version: 1. Access Date: 06.01.2020

[35] EUROPEAN PARLIAMENT AND COUNCIL OF THE EUROPEAN UNION: *concerning type-approval requirements for the general safety of motor vehicles, their trailers and systems, components and separate technical units intended therefore: Regulation (EC) Nr. 661/2009.* July 13th 2009

[36] EWIN, N. J.: *Traction Control for Electric Vehicles with Independently Driven Wheels*, University of Oxford, Dissertation, 2016

[37] FODOR, M. ; YESTER, J. ; HROVAT, D. : Active Control of Vehicle Dynamics. In: *17th Digital Avionics Systems Conference* (1998)

[38] FRITZ, M. ; BRÖDE, P. ; FISCHER, S. : Assessment of Vibration-Induced Low Back Disorders - Comparison of the Vibration Evaluation According to VDI 2057 with a Force-Related Evaluation. In: *Zentralblatt für Arbeitsmedizin, Arbeitsschutz und Ergonomie* 53 (2003), Nr. 6, pages 354–363

[39] FUJII, K. ; FUJIMOTO, H. : Traction Control based on Slip Ratio Estimation without Detecting Vehicle Speed for Electric Vehicle. In: *IEEE Power Conversion Conference* (2007)

[40] FUJIMOTO, H. ; AMADA, J. ; MAEDA, K. : Review of Traction and Braking Control for Electric Vehicle. In: *IEEE Power and Propulsion Conference* (2012), pages 1292–1299

[41] FUJIMOTO, H. ; FUJII, K. ; TAKAHASHI, N. : Traction and Yaw-Rate Control of Electric Vehicle with Slip-Ratio and Cornering Stiffness Estimation. In: *Proceedings of the American Control Conference* (2007), pages 5742–5747

[42] GERHARD, J. ; HÖNNINGER, H. ; BISCHOF, H. : A New Approach to Functional and Software Structure for Engine Management Systems - BOSCH ME7. In: *SAE Transactions - Journal of Engines* (1998)

[43] GILLESPIE, T. D.: *Fundamentals of Vehicle Dynamics*. 4th printing. Warrendale, PA : Society of Automotive Engineers, 1992. – ISBN 978–1560911999

[44] GÖTTING, G. : *Dynamische Antriebsregelung von Elektrostraßenfahrzeugen unter Berücksichtigung eines schwingungsfähigen Antriebsstrangs*. Aachen, Germany, Rheinisch Westfälische Technische Hochschule, Dissertation, 2004

[45] GUO, H. ; YU, R. ; QIANG, W. ; CHEN, H. : Optimal Slip Based Traction Control for Electric Vehicles Using Feedback Linearization. In: *International Conference on Mechatronics and Control (ICMC)* (2014), pages 1159–1164

[46] GUSTAFSSON, F. ; AHLQVIST, S. ; FORSSELL, U. ; PERSSON, N. : Sensor Fusion for Accurate Computation of Yaw Rate and Absolute Velocity. In: *SAE World Congress* (2001), Nr. 2001-1-1064

[47] GUZZELLA, L. ; ONDER, C. H.: *Introduction to Modeling and Control of Internal Combustion Engine Systems*. 2. Aufl. s.l. : Springer-Verlag, 2010. – ISBN 978–3–642–10774–0

[48] HAMZAH, N. ; ARIPIN, M. K. ; SAM, Y. M. ; SELAMAT, H. ; GHAZALI, R. : Second Order Sliding Mode Controller for Longitudinal Wheel Slip Control. In: *IEEE International Colloqium on Signal Processing and its Applications* 8th (2012)

[49] HAO, N. ; GUO, H. ; CHEN, H. : Braking Controller Design for Electrical Vihicle Using Feedback Linearization Method. In: *Proceedings of the Chinese Control Conference* 34th (2015), pages 8085–8090

[50] HARAFI, A. ; AGHAGOLZADEH, A. ; ALLIZADEH, G. ; SADEGHI, M. : Designing a Sliding Mode Controller for Slip Control of Antilock Brake Systems. In: *Transportation Research Part C: Emerging Technologies* 16 (2008), Nr. 6, pages 731–741

[51] HÄRKEGARD, O. : *Backstepping and Control Allocation with Applications to Flight Control*, Linköping University, Dissertation, 2003

[52] HÄRKEGARD, O. : Dynamic Control Allocation Using Constrained Quadratc Programming. In: *Journal of Guidance, Control and Dynamics* 27 (2004), Nr. 6, pages 1028–1034

[53] HÄRKEGARD, O. ; GLAD, T. : Resolving Actuator Redundancy - Optimal Control vs. Control Allocation. In: *automatica* 41 (2005), pages 137–144

[54] HE, H. ; PENG, J. ; XIONG, R. ; FAN, H. : An Acceleration Slip Regulation Strategy for Four-Wheel Drive Electric Vehicles Based on Sliding Mode Control. In: *Energies* 7 (2014), Nr. 6, pages 3748–3763

[55] HEISSING, B. (Hrsg.) ; ERSOY, M. (Hrsg.) ; GIES, S. (Hrsg.): *Fahrwerkhandbuch: Grundlagen · Fahrdynamik · Komponenten · Systeme · Mechatronik · Perspektiven*. 4., überarb. u. erg. Aufl. 2013. Springer Fachmedien Wiesbaden (ATZ / MTZ-Fachbuch). http://dx.doi.org/10.1007/978-3-658-01992-1. http://dx.doi.org/10.1007/978-3-658-01992-1. – ISBN 978–3–658–01992–1

[56] HOFMANN, P. : *Hybridfahrzeuge: Ein alternatives Antriebssystem für die Zukunft*. 2. Aufl. Springer. http://dx.doi.org/10.1007/978-3-7091-1780-4. http://dx.doi.org/10.1007/978-3-7091-1780-4. – ISBN 978–3–7091–1780–4

[57] HORI, Y. ; TOYODA, Y. ; TSURUOKA, Y. : Traction Control of Electric Vehicle: Basic Experimental Results Using the Test EV UOT Electric March. In: *IEEE Transactions on Industy Applications* 34 (1998), Nr. 5

[58] HROVAT, D. ; FODOR, M. : Automotive Engine-Based Traction Control. In: *The Impact of Control Technology* 2nd (2014)

[59] HU, J.-S. ; YIN, D. ; HORI, Y. ; HU, F.-R. : A New MTTE Methology for Electric Vehicle Traction Control. International Conference on Electrical Machines and Systems (2009)

[60] HU, T. ; LIN, Z. ; CHEN, B. M.: An Analsis and Design Method for Linear Systems Subject to Actuator Saturation and Disturbance. Automatica (2002), Nr. 38, pages 351–359

[61] ISERMANN, R. : *Identifikation dynamischer Systeme 1: Grundlegende Methoden.* Zweite neubearbeitete und erweiterte Auflage. Springer (Springer-Lehrbuch). http://dx.doi.org/10.1007/978-3-642-84679-3. http://dx.doi.org/10.1007/978-3-642-84679-3. – ISBN 978–3–642–84679–3

[62] ISERMANN, R. : *Fahrdynamik-Regelung: Modellbildung, Fahrerassistenzsysteme, Mechatronik ; mit 28 Tabellen.* 1. Aufl. Friedr. Vieweg & Sohn Verlag — GWV Fachverlage GmbH Wiesbaden (Kraftfahrzeugtechnik). http://dx.doi.org/10.1007/978-3-8348-9049-8. http://dx.doi.org/10.1007/978-3-8348-9049-8. – ISBN 978–3–8348–9049–8

[63] ISIDORI, A. : *Nonlinear control systems.* Third edition, softcover reprint of the hardcover 3rd edition 2000. London : Springer, 1995 (Communications and control engineering). – ISBN 978–1–4471–3909–6

[64] IVANOV, V. : *Advanced Automotive Active Safety Systems: Focus on Integrated Chassis Control for Conventional and Electric Vehicles with Identification of Road Conditions.* Ilmenau, Germany, Technical University Ilmenau, Habilitation Thesis, 2017

[65] IVANOV, V. ; SAVITSKI, D. ; SHYROKAU, B. : A Survey of Traction Control and Anti-lock Braking Systems of Full Electric Vehicles with Individually-Controlled Electric Motors. In: *IEEE Transactions on Vehicluar Technology* 64 (2015), Nr. 9, pages 3878–3896

[66] JAIME, R.-P. ; KÄSTNER, F. ; ERBAN, A. ; KNÖDLER, K. : Regelalgorithmen für Rekuperation und Traktion bei Elektrofahrzeugen. In: *ATZ elektronik* 02 (2014), pages 12–17

[67] JOHANSEN, T. A. ; FOSSEN, T. I.: Control Allocation - A Survey. In: *automatica* 49 (2013), Nr. 5, pages 1087–1103

[68] JOHN, S. ; PEDRO, J. : Hybrid Feedback Linearization Slip Control for Anti-Lock Braking System. In: *Acta Polytechnica Hungaria* 10 (2013), pages 81–99

[69] KABGANIAN, M. ; KAZEMI, R. : A New Strategy for Traction Control in Turning via Engine Modeling. In: *IEEE Transactions on Vehicluar Technology* 50 (2001), Nr. 06, pages 1540–1548

[70] KHALIL, H. K.: *Nonlinear systems.* Pearson new internat. ed., 3. ed. Harlow : Pearson Education, 2014 (Always learning). – ISBN 978–1–292–03921–3

[71] KHATUN, P. ; BINGHAM, C. M. ; SCHOFIELD, N. ; MELLOR, P. H.: Application of Fuzzy Control Algorithms for Electric Vehicle Antilock Braking/Traction COntrol Systems. In: *IEEE Transactions on Vehicluar Technology* 52 (2003), Nr. 2, pages 1356–1364

[72] KILLIAN, D. ; FISCHER, S. ; LIENKAMP, M. ; POLTERSDORF, S. ; SCHWARZ, R. : Combined Control Strategy for the Combustion Engine and Brake System to Enhance the Driving Dynamics and Traction of Front-Wheel-Drive Vehicles. In: *International Munich Chassis Symposium* 6th (2015), pages 629–645

[73] KNOBEL, C. ; PRUCKNER, A. ; BÜNTE, T. : Optimized Force Allocation: A General Approach to Control and to Investigate the Motion of Over-Actuated Vehicles. In: *IFAC Proceedings* 39 (2006), Nr. 16, pages 366–371

[74] KRÜGER, J. ; PRUCKNER, A. ; KNOBEL, C. : Control Allocation for Road Vehicles: A System-Independent Approach for Integrated Vehicle Dynamics Control. In: *Aachener Kolloquium FAhrzeug- und Motorentechnik* 19th (2010)

[75] KUHN, U. : Eine praxisnahe Einstellregel fuer PID-Regler: Die T-Summen-Regel. In: *Automatisierungstechnische Praxis* 37 (1995), Nr. 5, pages 10–20

[76] LAGARIAS, J. ; REEDS, J. ; WRIGHT, M. ; WRIGHT, P. : Convergence Properties of the Nelder Mead Simplex Method in Low Dimensions. In: *SIAM Journal of Optimization* 9 (1998), Nr. 4, pages 112–147

[77] LASCHET, A. : *Fachberichte Simulation*. Bd. 9: *Simulation von Antriebssystemen: Modellbildung der Schwingungssysteme und Beispiele aus der Antriebstechnik*. Springer. http://dx.doi.org/10.1007/978-3-642-83531-5. http://dx.doi.org/10.1007/978-3-642-83531-5. – ISBN 978–3–642–83531–5

[78] LENG, B. ; XIONG, L. ; YOU, Z. ; ZOU, T. : Allocation Control Algorithms Design and Comparison Based on Distributed Drive Electric Vehicles. In: *International Journal of Automotive Technology* 19 (2018), Nr. 1, pages 55–62

[79] LI, S. ; NAKAMURA, K. ; KAWABE, T. ; MORIKAWA, K. : A Sliding Mode Control for Slip Ratio of Electric Vehicle. In: *Proceedings of Society of Instrument and Control Engineers Annual Conference (SICE)* (2012)

[80] LIANG, B.-R. ; LIN, W.-S. : A New Slip Ratio Observer and its Applicatio in Electric Vehicle Wheel Slip Control. In: *IEEE International Conference on Systems, Man and Cybernetics (SMC)* (2012), pages 41–46

[81] LITZ, L. : *Hochschulsammlung Ingenieurwissenschaft. Datenverarbeitung*. Bd. 4: *Reduktion der Ordnung linearer Zustandsraummodelle mittels modaler Verfahren*. Stuttgart : Hochschulverl., 1979. – ISBN 978–3810720580

[82] LOOF, J. W. ; BESSELINK, I. J. M. ; NIJMEIJER, H. : Traction control of an electric formula student racing car. In: *Proceedings of the FISITA 2014 World Automotive Congress* (2014)

[83] LOOMAN, C. : *Analyse, Modellierung und Entwurf einer Mehrgrößenregelung zur aktiven Schwingungsdämpfung im hybriden Antriebsstrang*. Dresden, Technische Universität Dresden, Dissertation, 2012

[84] LU, H. ; LI, J. ; ZHOU, B. ; GAO, Y. ; SUN, B. : Traction and Yaw-Rate Control of 4WD Vehicle with Drive Force Distribution. In: *IEEE International Conference on Electronic & Mechanical Engineering and Information Technology* (2011), pages 1062–1065

[85] LUENBERGER, D. G. ; YE, Y. : *International series in operations research & management science*. Bd. 228: *Linear and nonlinear programming*. 4. ed. Cham : Springer, 2016. – ISBN 978–3–319–18841–6

[86] LUNZE, J. : *Regelungstechnik*. 8., neu bearb. Aufl., 2010. Berlin : Springer, 2010 (Springer-Lehrbuch). – ISBN 978–3–642–13807–2

[87] LUNZE, J. : *Regelungstechnik 2: Mehrgrößensysteme, digitale Regelung*. 6., neu bearbeitete Aufl., 2010. Springer (Springer-Lehrbuch). http://dx.doi.org/10.1007/978-3-642-10198-4. http://dx.doi.org/10.1007/978-3-642-10198-4. – ISBN 978–3–642–10197–7

[88] MAEDA, K. ; FUJIMOTO, H. ; HORI, Y. : Four-Wheel Driving-Force Distribution Method for Instantaneous or Split Slippery Roads for Electric Vehicle with In-Wheel Motors. In: *IEEE International Workshop on Advanced Motion Control (AMC)* 12th (2012)

[89] MERKER, G. P. ; TEICHMANN, R. : *Grundlagen Verbrennungsmotoren: Funktionsweise, Simulation, Messtechnik*. 7th. Wiesbaden : Springer Vieweg, 2014

[90] MICHALKA, A. : *Ein Modellbasiertes Gesamtkonzept zur Längsdynamiksteuerung und -regelung von Parallelhybridfahrzeugen*. Erlangen, Friedrich-Alexander-Universität, Dissertation, 2014

[91] MITSCHKE, M. ; WALLENTOWITZ, H. : *Dynamik der Kraftfahrzeuge*. 5., überarb. u. erg. Aufl. Springer Vieweg (VDI-Buch). http://dx.doi.org/10.1007/978-3-658-05068-9. http://dx.doi.org/10.1007/978-3-658-05068-9. – ISBN 978–3–658–05067–2

[92] MORAAL, P. ; KOLMANOVSKY, I. : Turbocharger Modeling for Automotive Control Applications. In: *SAE Transactions - Journal of Engines* 01 (1999), pages 1–15

[93] MÜLLER, G. ; GREGULL, V. ; BRÄSEMANN, C. ; MÜLLER, S. : Development of a Real-Time Friction Estimation Procedure. In: *Proceedings of chassis.tech* (2018), pages 249–263

[94] MÜLLER, G. ; LESCHIK, C. ; GREGULL, V. ; SIERON, N. ; MÜLLER, S. : *Unfallvermeidung durch Reibwertprognosen – Umsetzung und Anwendung*. Bd. Forschungsvereinigung Automobiltechnik e.V. (FAT). FAT-Schriftenreihe 322. Berlin : Forschungsvereinigung Automobiltechnik e.V. (FAT), 2019

[95] NAKAKUKI, T. ; SHEN, T. ; TAMURA, K. : Adaptive Control Approach to Uncertain Longitudinal Tire Slip in Traction Control of Vehicles. In: *Asian Journal of Control* 10 (2008), Nr. 1, pages 67–73

[96] NELDER, J. A. ; MEAD, R. : A Simplex Method for Function Minimization. In: *The Computer Journal* 7 (1965), Nr. 4, pages 308–313

[97] NYANDORO, O. ; PEDRO, J. ; DWOLATZKY, B. ; DAHUNSI, O. : State Feedback Based Linear Slip Control Formulation for Vehicular Antilock Braking System. In: *Proceedings of the World Congress on Engineering (WCE)* 1 (2011)

[98] PACEJKA, H. B. ; BESSELINK, I. : *Tire and vehicle dynamics*. 3rd ed. Elsevier/BH http://search.ebscohost.com/login.aspx?direct=true&scope=site&db=nlebk&db=nlabk&AN=453741. – ISBN 978–0–08–097016–5

[99] PAPAGEORGIOU, M. ; LEIBOLD, M. ; BUSS, M. : *Optimierung: Statische, dynamische, stochastische Verfahren für die Anwendung.* 3., neu bearb. u. erg. Aufl. 2012. Springer. http://dx.doi.org/10.1007/978-3-540-34013-3. http://dx.doi.org/10.1007/978-3-540-34013-3. – ISBN 978–3–540–34012–6

[100] PARK, J. ; JEONG, H. ; JANG, I. G. ; HWANG, S. H.: Torque Distribution Algorithm for an Independently Driven Electric Vehicle Using a Fuzzy Control Method. In: *Energies* 8 (2015), Nr. 8, pages 8537–8551

[101] PHAM, T. ; BUSHNELL, L. : Two-Degree-of-Freedom Damping Control of Driveline Oscillations caused by Pedal Tip-In Maneuver. In: *American Control Conference* (2015), pages 1425–1432

[102] PHAM, T. ; SEIFRIED, R. ; HOCK, A. ; SCHOLZ, C. : Nonlinear Flatness-Based Control of Driveline Oscillations for a Powertrain with Backlash Traversing. In: *IFAC Symposium on Advances in Automotive Control* 49 (2016), Nr. 11, pages 749–755

[103] PINTO, S. ; CHATZIKOMIS, C. ; SORNIOTTI, A. ; MANTRIOTA, G. : Comparison of Traction Controllers for Electric Vehicles With On-Board Drivetrains. In: *IEEE Vehicular Technology Society* 66 (2017), Nr. 8, pages 6715–6727

[104] RAJAMANI, R. : *Vehicle Dynamics and Control.* Rajesh Rajamani (Mechanical Engineering Series). http://dx.doi.org/10.1007/978-1-4614-1433-9. http://dx.doi.org/10.1007/978-1-4614-1433-9. – ISBN 978–1–4614–1433–9

[105] REICHENSDÖRFER, E. ; ODENTHAL, D. ; WOLLHERR, D. : On the Stability of Nonlinear Wheel-Slip Zero Dynamics in Traction Control Systems. In: *IEEE Transactions on Control Systems Technology* PP (2018), pages 1–16

[106] REIF: *Automobilelektronik.* Springer Fachmedien Wiesbaden, 2014. – ISBN 978–3–658–05047–4

[107] REIF, K. (Hrsg.): *Automotive Mechatronics: Automotive Networking, Driving Stability Systems, Electronics.* Springer Fachmedien Wiesbaden (Bosch Professional Automotive Information). http://dx.doi.org/10.1007/978-3-658-03975-2. http://dx.doi.org/10.1007/978-3-658-03975-2. – ISBN 978–3–658–03975–2

[108] REIF, K. ; NOREIKAT, K.-E. ; BORGEEST, K. : *Kraftfahrzeug-Hybridantriebe: Grundlagen, Komponenten, Systeme, Anwendungen.* Vieweg+Teubner Verlag (ATZ / MTZ-Fachbuch). http://dx.doi.org/10.1007/978-3-8348-2050-1. http://dx.doi.org/10.1007/978-3-8348-2050-1. – ISBN 978–3–8348–2050–1

[109] RILL, G. : Wheel Dynamics. In: *Proceedings of the XII International Symposium on Dynamic Problems of Mechanics (DINAME 2007)* (2007), Nr. Brasília, Brazil

[110] ROSENBERGER, M. ; KIRSCHNECK, M. ; KOCH, T. ; LIENKAMP, M. : Hybrid-ABS. Integration der elektrischen Antriebsmotoren in die ABS-Regelung. In: *Automobiltechnisches Kolloquium* (2011)

[111] ROSENBERGER, M. ; SCHINDELE, F. ; KOCH, T. ; LIENKAMP, M. : Analyse und aktive Dämpfung von Antriebsstrangschwingungen bei Elektrofahrzeugen während der ABS-Regelung. In: *Tag des Fahrwerks* 8th (2012)

[112] ROSENBERGER, M. ; UHLIG, R. A. ; KOCH, T. ; LIENKAMP, M. : Combining Regenerative Braking and Anti-Lock Braking for Enhanced Braking Performance and Efficiency. In: *SAE World Congress* (2012)

[113] SATZGER, C. ; CASTRO, R. de ; BÜNTE, T. : A Model Predictive Control Allocation Approach to Hybrid Braking of Electric Vehicles. In: *IEEE Intelligent Vehicles Symposium (IV)* (2014), pages 286–292

[114] SCHINKEL, M. ; HUNT, K. : Anti-Lock Braking Control using a Sliding Mode like Approach. In: *Proceedings of the American Control Conference* (2002), pages 2386–2391

[115] SCHNEIDER, S. : *Rechnergestützte, kooperativ arbeitende Optimierungsverfahren am Beispiel der Fabriksimulation*. Kassel, University of Kassel, Dissertation, 2001

[116] SCHRAMM, D. ; HILLER, M. ; BARDINI, R. : *Modellbildung und Simulation der Dynamik von Kraftfahrzeugen*. 2., vollständig überarbeitete Auflage. Springer Vieweg (SpringerLink). http://dx.doi.org/10.1007/978-3-642-33888-5. http://dx.doi.org/10.1007/978-3-642-33888-5. – ISBN 978–3–642–33887–8

[117] SCHRAMM, D. ; HILLER, M. ; BARDINI, R. : *Vehicle Dynamics: Modeling and Simulation*. 2nd ed. 2018. Springer Berlin Heidelberg. http://dx.doi.org/10.1007/978-3-662-54483-9. http://dx.doi.org/10.1007/978-3-662-54483-9. – ISBN 978–3–662–54482–2

[118] SCHRÖDER, D. : *Elektrische Antriebe - Grundlagen: Mit durchgerechneten Übungs- und Prüfungsaufgaben*. 5., erweiterte Auflage, 2013. Springer Berlin Heidelberg (Springer-Lehrbuch). http://dx.doi.org/10.1007/978-3-642-30471-2. http://dx.doi.org/10.1007/978-3-642-30471-2. – ISBN 978–3–642–30470–5

[119] SCHRÖDER, D. : *Elektrische Antriebe - Regelung von Antriebssystemen*. 4. Auflage, 2015. Springer Vieweg. http://dx.doi.org/10.1007/978-3-642-30096-7. http://dx.doi.org/10.1007/978-3-642-30096-7. – ISBN 978–3–642–30095–0

[120] SCHWEERS, T. F.: *Entwicklung eines Testverfahrens für Antriebsschlupf-Regelsysteme*. Aachen, Germany, Rheinisch Westfälische Technische Hochschule, Diss., 1997

[121] SCHWEERS, T. F. ; WALLENTOWITZ, H. ; ROETH, H. : Proposal for Testing and Rating of Traction and Stability Control Systems. In: *Vehicle System Dynamics* 26 (1996), Nr. 2, pages 117–126

[122] SHAMASH, Y. : Model Reduction Using the Routh Stability Criterion and the Padé Approximation Technique. In: *International Journal of Control* 21 (1975), Nr. 3, pages 475–484

[123] SHOUBO, L. ; CHENGLIN, L. ; SHANGLOU, C. ; LIFANG, W. : Traction Control of Hybrid Electirc Vehicle. In: *IEEE Vehicle Power and Propulsion Conference* (2009), pages 1535–1540

[124] SÖFFKER, D. ; YU, T.-J. ; MÜLLER, P. C.: State Estiation of Dynamical Systems with Nonlinearities by Using Proportional-Integral Observer. In: *International Journal of Systems Science* 26 (2007), Nr. 9, pages 1571–1582

[125] SONG, J.-B. ; BYUN, K.-S. : Throttle Actuator Control System for Vehicle Traction Control. In: *Mechatronics* 9 (1999), pages 477–495

[126] STELLET, J. E. ; GIESSLER, M. ; GAUTERIN, F. : Model-Based Traction Control for Electric Vehicles. In: *ATZ elektronik* 9 (2014), Nr. 02, pages 44–50

[127] STOICUTA, O. ; CAMPIAN, H. ; PANA, T. :   The Comparative Study of the Stability of the Vector Control Systems That Contain In the Loop Luenberger and Kalman Type Estimators. In: *Proceedings of IEEE Conference on Automation, Quality and Testing, Robotics, AQTR* (2006), pages 113–117

[128] SUBROTO, R. K.: Vehicle Stability Control of 4WD Electric Vehicle Using Combined Adaptive Sliding Mode Controller and Control Allocation Method. In: *IEEE International Future Energy Electronics Conference* 3rd (2017)

[129] SYRNIK, R. :   *Untersuchung der fahrdynamischen Potentiale eines elektromotorischen Traktionsantriebs.* Munich, Technical University of Munich, Dissertation, 2015

[130] TANELLI, M. ; PIRODDI, L. ; SAVARESI, S. M.:   Real-Time Identification of Tire-Road Friction Conditions. In: *IET Control Theory & Applications* 3 (2009), Nr. 7, pages 891–906

[131] VDI-RICHTLINIE 2057:   *Einwirkung mechanischer Schwingungen auf den Menschen.* 2017

[132] VON GRUNDHERR ZU ALTENTHAN UND WEIYHERHAUS, JOHANNES: *Ableitung einer heutristischen Betriebsstrategie für ein Hybridfahrezug aus einer Online-Optimierung.* Munich, Technical University of Munich, Dissertation, 2009

[133] WANG, H.-D. ; YI, J.-Q. ; FAN, G.-L. :   A Dynamic Control Allocation Method for Unmanned Aerial Vehicles with Multiple Control Effectors. In: *International Conference on Intelligent Robotics and Applications (ICIRA)* 1st (2008), pages 1175–1184

[134] WEBERSINKE, L. :   *Adaptive Antriebsstrangregelung für die Optimierung des Fahrverhaltens von Nutzfahrzeugen.* Karlsruhe, Universität Karlsruhe, Dissertation, 2008

[135] YEAP, K. Z. ; MÜLLER, S. :   Understanding the Influence of Traction on the Eigen Behaviour of Torsional Vibrations in Individual-Wheel Drives. In: *at - Automatisierungstechnik* 63 (2015), Nr. 6, pages 465–475

[136] YEAP, K. Z. ; MÜLLER, S. :   Characterizing the Interaction of Individual-Wheel Drives with Traction by Linear Parameter-Varying Model: A Method for Analyzing the ROle of Traction in Torsional Vibrations in Wheel Drives and Active Damping. In: *Vehicle System Dynamics* 54 (2016), Nr. 4, pages 258–280

[137] YIN, D. ; HORI, Y. :   Traction Control for EV Based on Maximum Transmissible Torque Estimation. In: *International Journal of Intelligent Transportation Systems Research* 8 (2010), Nr. 1, pages 1–9

[138] YIN, D. ; OH, S. ; HORI, Y. :   A Novel Traction Control for EV Based on Maximum Transmissible Torque Estimation. In: *IEEE Transactions on Industrial Electronice* 56 (2009), Nr. 6, pages 2086–2094

[139] ZACCARIAN, L. :   Dynamic Allocation for Input Redundant Control Systems. In: *automatica* 45 (2009), Nr. 6, pages 1431–1438

[140] ZECH, A. ; EBERL, T. ; MARX, C. ; MÜLLER, S. :   Analysis of the Potential of a New Control Approach for Traction Control Considering a P2-Hybrid Drivetrain. In: *Proceedings of International Munich Chassis Symposium* (2018), pages 324–353

[141] ZECH, A. ; EBERL, T. ; MÜLLER, S. : Analyse einer neuen kaskadierten Reglerstruktur für die Antriebsschlupfbegrenzung hochdynamischer Fahrzeugantriebe: Ansatz zur Dämpfung von Torsionsschwingungen im Antriebsstrang. In: *Proceedings of VDI/VDE Conference Autoreg* (2017), pages 65–76

[142] ZECH, A. ; EBERL, T. ; REICHENSDÖRFER, E. ; ODENTHAL, D. ; MÜLLER, S. : Method for Developing Tire Slip Controllers Regarding a New Cascaded Controller Structure. In: *Proceedings of International Symposium on Advanced Vehicle Control* (2018), pages 302–307

[143] ZEMKE, S. : *Analyse und modellbasierte Regleung von Ruckelschwingungen im Antriebsstrang von Kraftfahrzeugen.* Hannover, Gottfried Wilhelm Leibniz Universität, Dissertation, 2012

[144] ZHANG, A. ; HU, Q. ; HUO, X. : Dynamic Control Allocation for Spacecraft Attitude Stabilization with Actuator Uncertainty. In: *Journal of Aerospace Engineering* 228 (2014), Nr. 8, pages 1336–1347

[145] ZHANG, Y. ; ZHAO, Z. ; LU, T. ; YUAN, L. ; XU, W. ; ZHU, J. : A Comparative Study of Luenberger Observer, Sliding Mode Observer and Extended Kalman Filter for Sensorless Vector Control of Induction Motor Drives. In: *Proceedings of IEEE Energy Conversion Congress* (2009), pages 2466–2473

[146] ZHAO, F. ; LUO, Y. ; LI, K. : Traction Control Method for Hybrid Electric Vehicle Based on Multi-Objective Dynamic Coordination Control. In: *World Electric Vehicle Journal* 5 (2012), Nr. 2, pages 460–468

[147] ZHOU, L. ; XIONG, L. ; YU, Z. : A Research on Anti-Slip Regulation for 4WD Electriv Vehicle with In-Wheel Motors. In: *International Conference on Computer Science and Electronics Engineering (ICCSEE)* 2nd (2013)

Supervised Works of Students

[S1]　LIU, GUANGHUI : *Systemidentifikation des Fahrzeugantriebsstrangs und Entwurf robuster Regelungskonzepte für die Raddrehzahlregelung.* Master's Thesis at BMW AG, supervised by the Institute for Automation and Applied Informatics at Karlsruhe Institute of Technology, 2016.

[S2]　ENDL, MATTHIAS : *Entwicklung eines Simulationsmodells für ein 8-Gang Wandler Automatikgetriebe.* Bachelor's Thesis at BMW AG, supervised by the Faculty of Mechanical Engineering at Ostbayerische Technische Hochschule Regensburg, 2017.

[S3]　REIMUND, PATRICK : *Erstellung eines torsionsschwingungsfähigen Simulationsmodells eines Fahrzeugantriebsstrangs und Parametrierung eines G12 Hybridfahrzeugs.* Internship at BMW AG, 2017.

[S4]　MARX, CARSTEN : *Entwurf eines Momentenallokationsalgorithmus für die Antriebsschlupfbegrenzung eines parallelhybriden Antriebsstrangs.* Master's Thesis at BMW AG, supervised by the Faculty of Aerospace Engineering at University of Applied Sciences Aachen, 2018.

[S5]　BERCHTOLD, DOMINIK : *Parameteroptimierung für die Antriebsschlupfregelung.* Master's Thesis at BMW AG, supervised by the Chair of Electrical Drive Systems and Power Electronics at Technical University of Munich, 2018.

[S6]　HUNDEGGER, LINDA : *Inkonsistenzen bei der Auslegung von Reglern in der Fahrdynamikentwicklung.* Master's Thesis at BMW AG, supervised by the Laboratory for Product Development and Lightweight Design at Technical University of Munich, 2018.